T0137338

Stem Cells in Clinical Applications

Series Editor
Phuc Van Pham
Laboratory of Stem Cell Research & Application
University of Science, Vietnam National University
Ho Chi Minh City, Vietnam

Stem Cells in Clinical Applications brings some of the field's most renowned scientists and clinicians together with emerging talents and disseminates their cutting-edge clinical research to help shape future therapies. While each book tends to focus on regenerative medicine for a certain organ or system (e.g. Liver, Lung and Heart; Brain and Spinal Cord, etc.) each volume also deals with topics like the safety of stem cell transplantation, evidence for clinical applications, including effects and side effects, guidelines for clinical stem cell manipulation and much more. Volumes will also discuss mesenchymal stem cell transplantation in autoimmune disease treatment, stem cell gene therapy in pre-clinical and clinical contexts, clinical use of stem cells in neurological degenerative disease, and best practices for manufacturers in stem cell production. Later volumes will be devoted to safety, ethics and regulations, stem cell banking and treatment of cancer and genetic disease. This series provides insight not only into novel research in stem cells but also their clinical and real-world contexts. Each book in *Stem Cells in Clinical Applications* is an invaluable resource for advanced undergraduate students, graduate students, researchers and clinicians in Stem Cells, Tissue Engineering, Biomedical Engineering or Regenerative Medicine.

More information about this series at http://www.springer.com/series/14002

Phuc Van Pham
Editor

Stem Cell Drugs - A New Generation of Biopharmaceuticals

Editor
Phuc Van Pham
Laboratory of Stem Cell Research & Application
University of Science
Vietnam National University
Ho Chi Minh, Vietnam

ISSN 2365-4198 ISSN 2365-4201 (electronic)
Stem Cells in Clinical Applications
ISBN 978-3-030-07589-7 ISBN 978-3-319-99328-7 (eBook)
https://doi.org/10.1007/978-3-319-99328-7

This Springer imprint is published by the registered company Springer Nature Switzerland AG
The registered company address is: Gewerbestrasse 11, 6330 Cham, Switzerland

Preface

Stem cell therapy is a new therapy used in the treatment of various diseases. Since the first transplant of a primary product of stem cells in the 1950s, stem-cell-based products have a long history. After more than 60 years of development, stem-cell-based products can be grouped into six different generations, including stem-cell-enriched fractions (first generation), pure stem cells (second generation), long-term expanded allogeneic stem cells (third generation), genetically modified or differentiated stem cells (fourth generation), exosomes, extracellular vesicles, and stem cell extracts (fifth generation), and stem cells derived from tissues or organs (sixth generation). Since the third generation, stem-cell-based products used as drugs in the treatment of various diseases have been referred to as stem cell drugs.

To date, stem cell drugs of the third, fourth, and fifth generations are being used in clinics and commercially in several countries. Stem cell drugs have opened a new age of regenerative medicine. This book focuses on stem cell drugs of the third, fourth, and fifth generations of stem-cell-based products. In chapters 1–5, we introduce fifth-generation stem-cell-based products containing extracellular microvesicles obtained from stem cells. Chapters 1, 2, and 4 introduce some applications of microvesicles as cell-based, cell-free therapy in disease treatment and rejuvenation. Chapters 3 and 5 introduce some techniques to prepare and trigger microvesicle production from mesenchymal stem cells. In chapters 6–8, we focus on the third and fifth generations of stem-cell-based products. Chapter 6 introduces the evolution of stem cell products, while chapter 7 focuses on off-the-shelf mesenchymal stem cell technology. Chapter 8 presents some ethical and legal issues of cord blood stem cell banks.

In preparing this book, we aimed at making it accessible to not only those working in the field of stem cell biology, but also to nonexperts with a broad interest in stem cells and human health. We hope the book will be of value to all concerned with the new generation of stem-cell-based products, including stem cell drugs.

We are indebted to the authors who graciously accepted their assignments and who have infused the text with their energetic contributions. We are incredibly thankful to the staff at Springer for agreeing to publish the book.

Ho Chi Minh City, Vietnam Phuc Van Pham

Contents

Part I Microvesicles

1 **Using Stem Cell-Derived Microvesicles in Regenerative Medicine: A New Paradigm for Cell-Based-Cell-Free Therapy** 3
Mohammad Amin Rezvanfar, Mohammad Abdollahi, and Fakher Rahim

2 **Secretome: Pharmaceuticals for Cell-Free Regenerative Therapy** 17
Nazmul Haque, Basri Johan Jeet Abdullah, and Noor Hayaty Abu Kasim

3 **Preparation of Extracellular Vesicles from Mesenchymal Stem Cells** 37
Fernanda Ferreira Cruz, Ligia Lins de Castro, and Patricia Rieken Macedo Rocco

4 **Exosomes for Regeneration, Rejuvenation, and Repair** 53
Joydeep Basu and John W. Ludlow

5 **Proinflammatory Cytokines Significantly Stimulate Extracellular Vesicle Production by Adipose-Derived and Umbilical Cord-Derived Mesenchymal Stem Cells** 77
Phuc Van Pham, Ngoc Bich Vu, Khanh Hong-Thien Bui, and Liem Hieu Pham

Part II Stem Cells

6 **Evolution of Stem Cell Products in Medicine: Future of Off-the-Shelf Products** 93
Phuc Van Pham, Hoa Trong Nguyen, and Ngoc Bich Vu

7 **Off-the-Shelf Mesenchymal Stem Cell Technology** 119
Ngoc Bich Vu, Phuong Thi-Bich Le, Nhat Chau Truong, and Phuc Van Pham

8 **Ethical and Legal Issues of Cord Blood Stem Cell Banking** 143
Luciana Riva, Giovanna Floridia, and Carlo Petrini

Index 153

Contributors

Mohammad Abdollahi Department of Toxicology and Diseases, Pharmaceutical Sciences Research Center (PSRC), Tehran University of Medical Sciences (TUMS), Tehran, Iran

Department of Toxicology and Pharmacology, Faculty of Pharmacy, Tehran University of Medical Sciences (TUMS), Tehran, Iran

Basri Johan Jeet Abdullah Department of Biomedical Imaging, Faculty of Medicine, University of Malaya, Kuala Lumpur, Malaysia

Joydeep Basu Twin City Bio, LLC, Winston-Salem, NC, USA

Khanh Hong-Thien Bui University Medical Center, University of Medicine and Pharmacy, Ho Chi Minh City, Vietnam

Ligia Lins de Castro Laboratory of Pulmonary Investigation, Carlos Chagas Filho Biophysics Institute, Federal University of Rio de Janeiro, Rio de Janeiro, RJ, Brazil

Fernanda Ferreira Cruz Laboratory of Pulmonary Investigation, Carlos Chagas Filho Biophysics Institute, Federal University of Rio de Janeiro, Rio de Janeiro, RJ, Brazil

Giovanna Floridia Bioethics Unit, Istituto Superiore di Sanità (Italian National Institute of Health), Rome, Italy

Nazmul Haque Department of Restorative Dentistry, Faculty of Dentistry, University of Malaya, Kuala Lumpur, Malaysia

Department of Oral Biology and Biomedical Sciences, Faculty of Dentistry, MAHSA University, Selangor, Malaysia

Noor Hayaty Abu Kasim Department of Restorative Dentistry, Faculty of Dentistry, University of Malaya, Kuala Lumpur, Malaysia

Phuong Thi-Bich Le Van Hanh General Hospital, Ho Chi Minh City, Vietnam

John W. Ludlow Zen-Bio, Inc., Research Triangle Park, NC, USA

Hoa Trong Nguyen Stem Cell Institute, VNUHCM University of Science, Ho Chi Minh City, Vietnam

Carlo Petrini Bioethics Unit, Istituto Superiore di Sanità (Italian National Institute of Health), Rome, Italy

Liem Hieu Pham Pham Ngoc Thach University of Medicine, Ho Chi Minh City, Vietnam

Phuc Van Pham Laboratory of Stem Cell Research and Application, VNUHCM University of Science, Ho Chi Minh City, Vietnam

Stem Cell Institute, VNUHCM University of Science, Ho Chi Minh City, Vietnam

Faculty of Biology-Biotechnology, VNUHCM University of Science, Ho Chi Minh City, Vietnam

Fakher Rahim Health Research Institute, Research Center of Thalassemia and Hemoglobinopathies, Ahvaz Jundishapur University of Medical Sciences, Ahvaz, Iran

Mohammad Amin Rezvanfar Department of Toxicology and Diseases, Pharmaceutical Sciences Research Center (PSRC), Tehran University of Medical Sciences (TUMS), Tehran, Iran

Luciana Riva Bioethics Unit, Istituto Superiore di Sanità (Italian National Institute of Health), Rome, Italy

Patricia Rieken Macedo Rocco Laboratory of Pulmonary Investigation, Carlos Chagas Filho Biophysics Institute, Federal University of Rio de Janeiro, Rio de Janeiro, RJ, Brazil

Nhat Chau Truong Laboratory of Stem Cell Research and Application, VNUHCM University of Science, Ho Chi Minh City, Vietnam

Ngoc Bich Vu Laboratory of Stem Cell Research and Application, VNUHCM University of Science, Ho Chi Minh City, Vietnam

Stem Cell Institute, VNUHCM University of Science, Ho Chi Minh City, Vietnam

Part I
Microvesicles

Chapter 1
Using Stem Cell-Derived Microvesicles in Regenerative Medicine: A New Paradigm for Cell-Based-Cell-Free Therapy

Mohammad Amin Rezvanfar, Mohammad Abdollahi, and Fakher Rahim

Common treatments for various diseases are mainly a series of suppressing, modifying, or stimulating drugs that in addition to having unwanted side effects, in the long term with the advancement of the disease, lose their therapeutic efficacy to a large extent. Hence, the treatment of many diseases remains a major challenge in medical research. Among the promising therapeutic strategies that have been introduced in recent years, using mesenchymal stem cells has attracted significant attention. Stem cells are a kind of cells that have the ability to transform to all types of cells in the body. These cells have the ability to regenerate and differentiate into various types of cells, including blood, cardio, nervous, and cartilage cells; they can also be employed to repair various tissues of the body after injury and can be injected into some tissues, the most cells of which are destroyed such as intestine tissues. Transplanting and replacing damaged cells and repairing and fixing defects

M. A. Rezvanfar
Department of Toxicology and Diseases, Pharmaceutical Sciences Research Center (PSRC), Tehran University of Medical Sciences (TUMS), Tehran, Iran

M. Abdollahi
Department of Toxicology and Diseases, Pharmaceutical Sciences Research Center (PSRC), Tehran University of Medical Sciences (TUMS), Tehran, Iran

Department of Toxicology and Pharmacology, Faculty of Pharmacy, Tehran University of Medical Sciences (TUMS), Tehran, Iran

F. Rahim (✉)
Health Research Institute, Research Center of Thalassemia and Hemoglobinopathies, Ahvaz Jundishapur University of Medical Sciences, Ahvaz, Iran

P. V. Pham (ed.), *Stem Cell Drugs - A New Generation of Biopharmaceuticals*, Stem Cells in Clinical Applications, https://doi.org/10.1007/978-3-319-99328-7_1

in a damaged tissue is other abilities of the stem cells. Microvesicles (MVs) are integral components of the cell-to-cell communication network, which releasing from different cells have the ability to progressively become a center of attention in stem cell-based therapy.

Stem Cells

Stem cells are generally undifferentiated cells that have self-proliferative ability and are able to differentiate into specific cell lines (Tweedell 2017). Under certain physiological or laboratory conditions, these cells can be converted into cells with specific functions, such as muscle cells of the heart or insulin-producing cells in the pancreas (Ardeshiry Lajimi et al. 2013; Ebrahimi et al. 2014; Ebrahimi and Rahim 2014; Rahim et al. 2013; Saki et al. 2013; Shahrabi et al. 2014). Stem cells have two important properties that distinguish them from other cell types (Tweedell 2017). The first one is their regeneration ability; these cells are undifferentiated cells with the unlimited ability to reproduce. The second one is that they are capable of differentiating and producing any kind of cells in the body. Accordingly, these cells are classified in the three categories.

Neuronal precursor cells are multipotent progenitor stem cells with variable capacities; they have the ability to differentiate into neuronal cells, including neurons and oligodendrocytes. Evidence suggests that in patients with MS who have an effective myelin plasmosis, neuronal progenitor cells migrate to injury sites and participate in the repair of damaged tissue; in fact, due to the inherent lack of recovery process over time using exogenous neuroleptic precursor cells can dramatically enhance the capacity of central nervous system restoration (Podbielska et al. 2013). These cells are mainly isolated from adult adipose tissues and cultured in nonspecific culture media, which severely restricts their therapeutic use (Wankhade et al. 2016).

The other one are embryonic stem cells (ESCs) existing in the body of the embryo during the first weeks of its formation, which means these cells make up the body of the human embryo (Kugler et al. 2017). It is clear that these cells can form different types of tissues and organs. They are taken from the internal cell mass of the 14–16-day-old fetus and are able to make all the cells and tissues of a person.

The umbilical cord stem cells are other potent cells, which, like adult stem cells, can produce a variety of cells in the laboratory. There are two types of stem cells in the umbilical cord that are able to make blood, bone, and fat cells, and make a replacement for bone marrow cells in bone marrow transplantation (Broxmeyer 2011; Sideri et al. 2011). At birth, these cells can be removed by cutting the umbilical cord from the blood of umbilical veins. These cells are less capable of differentiating into tissues and organs than embryonic stem cells, but their differentiation is much easier. The cord matrix called Wharton's jelly is the source of adult mesenchymal stem cells.

Adult stem cells are undifferentiated cells that are found in various cells of human tissues and organs, and have the ability to regenerate and differentiate into a variety of specific cells of the body or organ (Yang et al. 2017). The initial roles of these cells in a living organism include the protection and repair of the tissues that are derived from it. Scientists have found adult stem cells in more tissues than they thought. These findings advised scientists to use these cells in transplant science. It is more than 30 years now that bone marrow from the transplant passes is used to separate stem cells from adult hematopoietic cells. Adult stem cells have been detached from many organs and tissues of the body, but the important thing is that there are very few of these cells in each tissue that reside in a particular area of that tissue for years. These hidden cells are activated with the advent of disease or tissue damage. The tissues containing adult stem cells include bone marrow, peripheral blood, brain, blood vessels, dental pulp, skeletal muscle, skin, liver, pancreas, cornea, retina, and digestive system.

Scientists in many laboratories are working to transform adult stem cells to specific types of cells to use them for the treatment of diseases and tissue damages. The therapeutic potential of these cells contribute to the replacement of dopamine-producing cells in the brain in Parkinson's disease, the production of insulin-like cells in diabetes (i.e., insulin-dependent diabetes), and the repair of degenerated muscle cells (Ballios and van der Kooy 2010; Barkho and Zhao 2011).

Therapeutic Use of Stem Cells

Among promising therapeutic strategies, stem cell transplantation strategy has been especially devoted to cure inflammatory responses and promote the regeneration of the central nervous system (CNS) (Aurora and Olson 2014). This therapeutic approach can be used as an effective tool to overcome existing disabilities to promote simultaneous myelin, neural cells, and suppression of harmful inflammatory responses. This means that exogenous stem cells can physically contribute to the regeneration of the CNS, or by triggering the trophic factors and mobilizing the topical precursor cells help promote the repair process of CNS injuries. On the other hand, due to their potential immune response properties, stem cells can play a role in suppressing progressive inflammatory responses in autoimmune diseases, such as MS. The candidate stem cells for the treatment of MS, include neural cells derived from neuroprotective cells (NPCs), embryonic stem cell (ES), and mesenchymal stem cell (MSCs) (Muraro et al. 2017; Sargent et al. 2017). In addition to the important role of stem cells in restoring and repairing tissue, they are used to treat various diseases, including defective ossification, brain damage, Parkinson's disease, heart attacks, and tendon rupture (Lunn et al. 2011). A urine-derived stem cell has been discovered with some applicable biological properties

(Kang et al. 2015). These stem cells can be found in humans and various animal species such as monkeys, pigs, and rabbits. The availability and low cost of these cells make them suitable for cell therapy. Clinical trials have shown that the transplanted uterine stem cells may be used to treat debilitating analgesic disorders, and possibly neurodegenerative diseases such as Parkinson, Huntington, and Alzheimer (Li et al. 2017).

Limitations on the Use of Stem Cells

In recent years, a new bunch of studies on stem cells have begun, with many advances and successes. However, there are still many problems that limit the therapeutic use of these cells (Choumerianou et al. 2008). In the term of bioethics, for example embryonic stem cells are derived from live fetuses, which is prohibited in many countries, because eliminating the fetus that is capable of becoming a human being is considered as the death of a human soul (Outka 2009). However, compared to embryonic stem cells, adult stem cells are taken from the adult body with no damage to the body; thereafter, the use of adult stem cells does not have such limitations. At the same time, other potentially and actual applications of the mentioned cells in the medical field are highly sought after in the rest of the world. Another issue is the rejection of stem cells by the body. Since adult stem cells can be used for their own treatment, after injection into the patient's body, the immune system does not consider these cells as alien cells. It is worth mentioning that rejection of stem cells by the body is one of the major constraints facing researchers in the use of embryonic stem cells, since the antigenicity of these cells is not the same as that of the receptor, thus their probability of resuscitation rises. Of course, research is underway to suppress the supplying molecules of antigens to resolve this problem (Cabrera et al. 2006). Unwanted differentiation is also should be considered in stem cell therapy. Embryonic stem cells have such a high reproducibility and differentiation potential that they sometimes spontaneously transform into other cells without any particular treatment. Therefore, they must be prevented from accidental and unwanted differentiation.

Mature stem cells also have a great tendency of reproducibility in culture. Therefore, they are subjected to special treatments in the direction of targeted differentiation. Therefore, one of the major problems with the proliferation and differentiation of stem cells is that the orientation and direction of the differentiation of these cells into other cells that is somewhat hard and unknown. Nevertheless, if the path of multiplication and differentiation is identified, the appearance of different mammalian cells during embryonic development can be recognized, and as a result, it will be possible to identify the genes involved in the development of various cells (such as the heart and nerves). Here, the advantage of embryonic stem cells over adult stem cells is that adult cells do not give us such information (Penna et al. 2015). Also, *arrhythmia occurs* when stem cells, especially embryonic stem cells are used to repair damaged heart tissue; in fact in some cases, there is an inconsistency between the original and the repaired tissue. This causes the discontinuity of these two parts and as a result the heart rate rhythm breaks down. An inconsistency has been seen in some of the experiments performed on mice (Tian et al. 2015). However, this problem does not come about in autologous adult stem cells received from the patient. Due to the above limitations, in recent years, scientists have been focused on indirect and healthier use of mesenchymal stem cells based on the use of exosomes derived from these cells. Exosomes are cell-mediated microsomal cells, through which many of the paracrine effects of cells are revealed. The efficacy of mesenchymal stem cell-derived exosomes has been proven to be in the process of repairing and reconstructing a wide range of empirical patterns of tissue damage, which can reflect the anti-inflammatory and regenerative profiles of mesenchymal stem cells (Yu et al. 2014). Generally, the use of exosomes as a noncellular treatment method is advantageous over cellular therapy. In summary, exosomes are more stable and structurally functional than cells, and have more unlimited storage capacity (Lai et al. 2010). In addition, the stimulatory or inhibitory signaling induced by these exosomes is much stronger than that of the cells (Farsad 2002). Studies show that exosome therapy can be considered as a new strategy to overcome the current limitations of cell therapy.

A Perspective on Stem Cells' Microvesicles

Stem cell-derived EVs are circular fragments of membrane released from the endosomal compartment as exosomes, which play an important role in the biological functions of their parental cells (Yin and Jiang 2015). It is believed that EVs may simulate the effects of supportive blood-forming of their parent cells. The proregenerative effects of EVS are due to enriched bioactive lipids, antiapoptotic and prostimulatory growth factors or cytokines, as well as they deliver mRNAs, regulatory miRNAs, and proteins that improve the overall cell function. Therefore, EVs may open novel perspectives in the field of tissue regeneration and repair. Besides, the use of EVs instead of stem cells could represent a safe and potentially more advantageous alternative to cell-therapy approaches. Researchers investigating the effect of leukemia EVs isolated from acute myeloid leukemia patients on hematopoietic stem cells, suggest that these EVs can induce some effects on hematopoietic stem cells such as promoting cell survival (Razmkhah et al. 2017). So far, many studies have tested the potential clinical and experimental use of stem cell-derived EVs (Table 1.1).

These studies mostly have used MSC-derived EVs, and, as far as the authors of the present study are concerned, little attempt is made in using other types of stem cell-derived EVs. It has been shown that MSC-derived EVs have the capacity to mitigate radiation injury to marrow stem cells, so it can reverse radiation damage to bone marrow stem cells (BMSCs) (Wen et al. 2016). Besides, MSC-derived EVs (especially hematopoiesis-supporting effects of their parent cells) play a crucial role in the biological functions, since these EVs containing microRNAs that are involved

Table 1.1 Available studies on stem cell-derived microvesicles used in the treatment of various diseases

Authors	Country	Stem cell type	Disorder	Findings
Ji et al. (2017)	China	hESC-MSCs	Leukemia cells	Inhibited tumor growth and stimulated autophagy and excessive autophagy might induce apoptosis.
Nargesi et al. (2017)	USA	MSC-EVs	Renal injury and dysfunction	Testing the efficacy of MSC-derived EVs for treating renal disease.
Moore et al. (2017)	UK	MSC-EVs	Various cancers	The use of immunotherapy in combination with the advent of EVs provides potent therapies to various cancers.
Jaimes et al. (2017)	Germany	MSC-EVs	Microglia cells	MSC-EVs might represent a modulator of microglia activation with future therapeutic impact.
Drommelschmidt et al. (2017)	Germany	MSC-EVs	Brain injury	MSC-EVs may serve as a novel therapeutic option by preventing neuronal cell death, restoration of white matter microstructure, reduction of gliosis and long-term functional improvement.
Riazifar et al. (2017)	USA	MSC-EVs	Injured tissues	EVs are considered as potential therapeutic alternatives to cells for clinical applications.
Liu et al. (2016)	USA	MSC-EVs	Rupture of intracranial aneurysm	Prevented the rupture of intracranial aneurysm, in part due to their anti-inflammatory effect on mast cells, which was mediated by PGE2 production and EP4 activation.
Xie et al. (2016a)	China	MSC-EVs	Alginate-polycaprolactone	This EVs-alginate-PCL construct may offer a novel, proangiogenic, and cost-effective option for bone tissue engineering.
Baulch et al. (2016)	USA	Human neural stem cells (hNSC-EVs)	Irradiated brain	Reduce inflammation and preserves the structural integrity of the irradiated microenvironment.
Xie et al. (2016b)	China	MSC-EVs	Ex vivo expansion	Offer a promising therapeutic approach in CB transplantation.

(continued)

Table 1.1 (continued)

Authors	Country	Stem cell type	Disorder	Findings
Monsel et al. (2016)	France	MSC-EVs	Acute lung injury and other inflammatory lung diseases	Require large-scale production and standardization concerning identification, characterization, and quantification.
Lopez-Verrilli et al. (2016)	Chile	Menstrual MSCs	Neuritic outgrowth	Potential use of MenSCs as therapeutic conveyors in neurodegenerative pathologies.
Xie et al. (Xie et al. 2016c)	China	MSC-EVs	Tissue repair and antitumor experiments	Potential clinical translational opportunities of spheroid MSCs and MSC-EVs were discussed.
Farber and Katsman (2016)	USA	mESC-EVs	Retinal regeneration	Induce these processes and change Müller cells' microenvironment toward a more permissive state for tissue regeneration.
Yin and Jiang (2015)	China	MSC-EVs	Regeneration of injured tissues	The use of EVs instead of stem cells could represent a safe and potentially more advantageous alternative to cell-therapy approaches.
Li et al. (2015)	China	MSC-exosomes	TISSUE REPAIR	Biofunction, paracellular transport, and treatment mechanism will help the transform to clinical application.
Wang et al. (2015)	China	BM-MSC-EVs	Renal fibrosis	Suggesting that these may play a role in the fibrosis of aging renal tissues.
Monsel et al. (2015)	France	MSCs	Severe pneumonia	Effective as the parent stem cells in severe bacterial pneumonia.
Bobis-Wozowicz et al. (2015)	Poland	hiPSC-EVs	Recipient mature heart	New concept of use of hiPSCs as a source of safe acellular bioactive derivatives for tissue regeneration.
Lin et al. (2014a)	China	BM-MSC-Evs	Glutamate injured PC12	Preliminary experimental and theoretical evidence for the use of BMMSC-EVs in the treatment of neural excited damage.
Chen et al. (2014)	China	MSC-EVs	Arterial hypertension	Produce similar beneficial effects for treating hypertension.

(continued)

Table 1.1 (continued)

Authors	Country	Stem cell type	Disorder	Findings
Favaro et al. (2014)	Italy	MSC-EVs	Type 1 diabetes	Can inhibit in vitro a proinflammatory response to an islet antigenic stimulus in type 1 diabetes.
Lin et al. (2014b)	China	rBM-MSC-EVs	Glutamate-induced injury	A promising strategy to treat cerebral injury or some other neuronal diseases involving excitotoxicity.
Raisi et al. (2014)	Iran	MSC-EVs	Sciatic nerve regeneration	Alternative for the improvement of rat sciatic nerve regeneration.
Mokarizadeh et al. (2013)	Iran	MSC-EVs	Sperm quality	Enhance quality parameters and adhesive properties of cryopreserved sperm following treatment with MSC-derived EVs.
Dorronsoro and Robbins (2013)	USA	hucMSCs-exosomes	Injured kidney	Easy to isolate and safer to use than the parental stem cells, could have significant clinical utility.
Bruno and Camussi (2013)	Italy	MSC-EVs	Tissue repair	EVs released from stem cells may deliver proteins, bioactive lipids, and nucleic acids to injured cells.
Camussi et al. (2013)	Italy	Stem cell	Paracrine action	EVs released from stem cells retain several biological activities that are able to reproduce the beneficial effects of stem cells in a variety of experimental models.
Katsman et al. (2012)	USA	ESC-EVs	Müller cells of retina	May turn on an early retinogenic program of differentiation.
Biancone et al. (2012)	Italy	MSC-EVs	Tissue repair	Offer novel therapeutic approaches in regenerative medicine to repair damaged tissues, as an alternative to stem cell-based therapy.
Fonsato et al. (2012)	Italy	HLSC-EVs	Hepatoma growth	Stem cells may inhibit tumor growth and stimulate apoptosis.
Mokarizadeh et al. (2012)	Iran	MSC-EVs	Tolerogenic signaling	MSC-derived EVs are potent organelles for the induction of peripheral tolerance and modulation of immune responses.

(continued)

Table 1.1 (continued)

Authors	Country	Stem cell type	Disorder	Findings
Herrera et al. (2010)	Italy	HLSC-EVs	Hepatectomized	Activate a proliferative program in remnant hepatocytes after hepatectomy by a horizontal transfer of specific mRNA subsets.
Bruno et al. (2009)	Italy	MSC-EVs	Tubular injury	Activate a proliferative program in surviving tubular cells after injury via a horizontal transfer of mRNA.
Ratajczak et al. (2006)	USA	ESC-EVs	Hematopoietic progenitors	Increase their pluripotency after horizontal transfer of ES-derived mRNA.

hESC-MSCs human embryonic stem cell derived-mesenchymal stem cells, *MSCs* mesenchymal stem cells, *BM-MSC-EVs* bone marrow mesenchymal stem cell-derived extracellular microvesicles, *rBM-MSC-EVs* rat bone marrow mesenchymal stem cell-derived extracellular microvesicles, *ESC-EVs* embryonic stem cell-derived extracellular microvesicles, *HLSC-EVs* human liver stem cell-derived microvesicles

in the regulation of hematopoiesis (Xie et al. 2016b). Moreover, it has been indicated that MSC-derived EVs have protective effects on glutamate injured PC12 cells; this may elucidate their mechanism of the neural damage repair, and introduce them as potential candidates for the treatment of neurological diseases (Lin et al. 2014a). There are some conditions under which MSC releases EVs; one of them is hypoxia that can improve the release of EVs from MSC, and may provide an appropriate condition for EVs harvesting (Bi et al. 2014). BM-MSC-derived EVs play a protective role in acute pancreatitis by reducing the level of preinflammatory cytokines and NFκBp65 nuclear displacement regulation, and can be used as a strategy for the treatment of severe acute pancreatitis induced by sodium thrombolytic as well (Yin et al. 2016).

MSCs have been shown to support the specific features of hematopoietic progenitor stem cells (HPSCs) in the hematopoietic microenvironment of the bone marrow. MSCs have been used in coexisting systems as a feeding layer for cord blood ex vivo proliferation to increase the relatively low number of umbilical cord blood stem cells and precursors. A study showed that MSC-derived EVs contain micro-RNAs that are involved in the regulation of hematopoiesis. They also showed that MSC-derived EVs can enhance the proliferation of single-core cells and cord blood-derived CD34+ cells and produce more primary precursor cells in vitro. In addition, when MSC-derived EVs are added to the umbilical-derived stem cell, they are able to improve the hematopoietic-supporting effects of MSCs. These findings emphasize the role of MSC-derived EVs in ex vivo cord blood proliferation and may offer promising therapeutic approaches in umbilical cord blood transplantation (Xie et al. 2016b). Tumor cell-derived EVs are considered as a pivotal mechanism of donor cells in various cancers. Numerous studies suggested that EVs released from tumor cells are involved in pathological regulation of bone cell formation in the metastatic site. This further strengthens the role of tumor cell-derived microvesicles in cancer progression and disease aggressiveness (Karlsson et al. 2016; Razmkhah et al. 2015; Zhu et al. 2014). Since the in vitro maintenance of pluripotency and undifferentiated propagation of embryonic stem cells (ESCs) needs close-fitting cell–cell interactions and effective intercellular signaling, researchers attempt to show that ESC-derived EVs may express stem cell-specific molecules, which may support self-renewal and expansion of adult stem cells (Ratajczak et al. 2006).

Conclusion and Future Perspectives

Despite all of these considerations, a more specific expression of the efficacy of exosome therapy and its differences with cell therapy require more time and more accurate monitoring. Contrary to numerous studies that have shown the effective justification of the long-term stem cell therapy, the fact that the effects of exosomes are stable is not yet clear.

The results of this chapter confirm that stem cell-derived EVs as effective biological modulators can be used in the treatment of many diseases, including autoimmune disorders. The findings suggest that MSC-EVs play an important role in the biological functions of their parental cells.

The possibility of frequent withdrawal from long-term cell cultures and using existing commercial compounds, easy and short separation time without the need for advanced laboratory equipments, high biosecurity, unlimited storage capability and allogeneic application efficiency are among the broad therapeutic advantages of stem cell derived EVs and exosomes.

Stem cell-derived EVs have the capability to change the cell phenotype and fate of other different cell populations. This capacity has been confirmed with numerous diverse cell and tissue combinations. There is a great potential for stem cell-derived EVs modulation in the tissue renewal or cell growth era. Furthermore, stem cell-derived EVs may be applied as appropriate diagnostic biomarkers in various diseases, as they are one of the best biomimetic nanocarriers for a variety of molecules, including nucleic acids, proteins, and chemicals.

Although EVs therapy may offer a novel and extremely exciting therapeutic strategy, some important aspects are yet to be considered before their clinical applications. Firstly, the large-scale culture of stem cells and extraction, purification, and GMP-based production of EVs (nucleic acids, lipids, and proteins) should be defined in detail. Secondly, their long-term safety, efficacy, stability, and biodistribution at different preparations/concentrations should be evaluated accurately. Hence, our knowledge of the MSC secretome is not enough and mainly based on in vitro studies, it is critically important to characterize the MSC-secreted factors in vivo, using more sensitive techniques to analyze their qualitative and quantitative changes in response to the cellular damage.

References

Ardeshiry Lajimi A, Hagh MF, Saki N, Mortaz E, Soleimani M, Rahim F (2013) Feasibility of cell therapy in multiple sclerosis: a systematic review of 83 studies. Int J Hematol Oncol Stem Cell Res 7(1):15–33

Aurora AB, Olson EN (2014) Immune modulation of stem cells and regeneration. Cell Stem Cell 15(1):14–25

Ballios BG, van der Kooy D (2010) Biology and therapeutic potential of adult retinal stem cells. Can J Ophthalmol 45(4):342–351

Barkho BZ, Zhao X (2011) Adult neural stem cells: response to stroke injury and potential for therapeutic applications. Curr Stem Cell Res Ther 6(4):327–338

Baulch JE, Acharya MM, Allen BD, Ru N, Chmielewski NN, Martirosian V, Giedzinski E, Syage A, Park AL, Benke SN et al (2016) Cranial grafting of stem cell-derived microvesicles improves cognition and reduces neuropathology in the irradiated brain. Proc Natl Acad Sci U S A 113(17):4836–4841

Bi XY, Huang S, Chen JL, Wang F, Wang Y, Guo ZK (2014) [Exploration of conditions for releasing microvesicle from human bone marrow mesenchymal stem cells]. Zhongguo Shi Yan Xue Ye Xue Za Zhi 22(2):491–495

Biancone L, Bruno S, Deregibus MC, Tetta C, Camussi G (2012) Therapeutic potential of mesenchymal stem cell-derived microvesicles. Nephrol Dial Transplant 27(8):3037–3042

Bobis-Wozowicz S, Kmiotek K, Sekula M, Kedracka-Krok S, Kamycka E, Adamiak M, Jankowska U, Madetko-Talowska A, Sarna M, Bik-Multanowski M et al (2015) Human induced pluripotent stem cell-derived microvesicles transmit RNAs and proteins to recipient mature heart cells modulating cell fate and behavior. Stem Cells (Dayton, Ohio) 33(9):2748–2761

Broxmeyer HE (2011) Insights into the biology of cord blood stem/progenitor cells. Cell Prolif 44(Suppl 1):55–59

Bruno S, Camussi G (2013) Role of mesenchymal stem cell-derived microvesicles in tissue repair. Pediatr Nephrol (Berlin, Germany) 28(12):2249–2254

Bruno S, Grange C, Deregibus MC, Calogero RA, Saviozzi S, Collino F, Morando L, Busca A, Falda M, Bussolati B et al (2009) Mesenchymal stem cell-derived microvesicles protect against acute tubular injury. J Am Soc Nephrol 20(5):1053–1067

Cabrera CM, Cobo F, Nieto A, Concha A (2006) Strategies for preventing immunologic rejection of transplanted human embryonic stem cells. Cytotherapy 8(5):517–518

Camussi G, Deregibus MC, Cantaluppi V (2013) Role of stem-cell-derived microvesicles in the paracrine action of stem cells. Biochem Soc Trans 41(1):283–287

Chen JY, An R, Liu ZJ, Wang JJ, Chen SZ, Hong MM, Liu JH, Xiao MY, Chen YF (2014) Therapeutic effects of mesenchymal stem cell-derived microvesicles on pulmonary arterial hypertension in rats. Acta Pharmacol Sin 35(9):1121–1128

Choumerianou DM, Dimitriou H, Kalmanti M (2008) Stem cells: promises versus limitations. Tissue Eng Part B Rev 14(1):53–60

Dorronsoro A, Robbins PD (2013) Regenerating the injured kidney with human umbilical cord mesenchymal stem cell-derived exosomes. Stem Cell Res Ther 4(2):39

Drommelschmidt K, Serdar M, Bendix I, Herz J, Bertling F, Prager S, Keller M, Ludwig AK, Duhan V, Radtke S et al (2017) Mesenchymal stem cell-derived extracellular vesicles ameliorate inflammation-induced preterm brain injury. Brain Behav Immun 60:220–232

Ebrahimi A, Rahim F (2014) Recent immunomodulatory strategies in transplantation. Immunol Investig 43(8):829–837

Ebrahimi A, Hosseini SA, Rahim F (2014) Immunosuppressive therapy in allograft transplantation: from novel insights and strategies to tolerance and challenges. Cent Eur J Immunol 39(3):400–409

Farber DB, Katsman D (2016) Embryonic stem cell-derived microvesicles: could they be used for retinal regeneration? Adv Exp Med Biol 854:563–569

Farsad K (2002) Exosomes: novel organelles implicated in immunomodulation and apoptosis. Yale J Biol Med 75(2):95–101

Favaro E, Carpanetto A, Lamorte S, Fusco A, Caorsi C, Deregibus MC, Bruno S, Amoroso A, Giovarelli M, Porta M et al (2014) Human mesenchymal stem cell-derived microvesicles modulate T cell response to islet antigen glutamic acid decarboxylase in patients with type 1 diabetes. Diabetologia 57(8):1664–1673

Fonsato V, Collino F, Herrera MB, Cavallari C, Deregibus MC, Cisterna B, Bruno S, Romagnoli R, Salizzoni M, Tetta C et al (2012) Human liver stem cell-derived microvesicles inhibit hepatoma growth in SCID mice by delivering antitumor microRNAs. Stem Cells (Dayton, Ohio) 30(9):1985–1998

Herrera MB, Fonsato V, Gatti S, Deregibus MC, Sordi A, Cantarella D, Calogero R, Bussolati B, Tetta C, Camussi G (2010) Human liver stem cell-derived microvesicles accelerate hepatic regeneration in hepatectomized rats. J Cell Mol Med 14(6b):1605–1618

Jaimes Y, Naaldijk Y, Wenk K, Leovsky C, Emmrich F (2017) Mesenchymal stem cell-derived microvesicles modulate lipopolysaccharides-induced inflammatory responses to microglia cells. Int J Cancer 35(3):812–823

Ji Y, Ma Y, Chen X, Ji X, Gao J, Zhang L, Ye K, Qiao F, Dai Y, Wang H et al (2017) Microvesicles released from human embryonic stem cell derived-mesenchymal stem cells inhibit proliferation of leukemia cells. Oncol Rep 38(2):1013–1020

Kang HS, Choi SH, Kim BS, Choi JY, Park G-B, Kwon TG, Chun SY (2015) Advanced properties of urine derived stem cells compared to adipose tissue derived stem cells in terms of cell proliferation, immune modulation and multi differentiation. J Korean Med Sci 30(12):1764–1776

Karlsson T, Lundholm M, Widmark A, Persson E (2016) Tumor cell-derived exosomes from the prostate cancer cell line TRAMP-C1 impair osteoclast formation and differentiation. PLoS One 11(11):e0166284

Katsman D, Stackpole EJ, Domin DR, Farber DB (2012) Embryonic stem cell-derived microvesicles induce gene expression changes in Muller cells of the retina. PLoS One 7(11):e50417

Kugler J, Huhse B, Tralau T, Luch A (2017) Embryonic stem cells and the next generation of developmental toxicity testing. Expert Opin Drug Metab Toxicol 13(8):833–841

Lai RC, Arslan F, Lee MM, Sze NS, Choo A, Chen TS, Salto-Tellez M, Timmers L, Lee CN, El Oakley RM et al (2010) Exosome secreted by MSC reduces myocardial ischemia/reperfusion injury. Stem Cell Res 4(3):214–222

Li X, Liu L, Chai J (2015) [Progress of mesenchymal stem cell-derived exosomes in tissue repair]. Zhongguo Xiu Fu Chong Jian Wai Ke Za Zhi, 29(2):234–238

Li J, Luo H, Dong X, Liu Q, Wu C, Zhang T, Hu X, Zhang Y, Song B, Li L (2017) Therapeutic effect of urine-derived stem cells for protamine/lipopolysaccharide-induced interstitial cystitis in a rat model. Stem Cell Res Ther 8(1):107

Lin SS, Zhu B, Guo ZK, Huang GZ (2014a) [Protective effect of bone marrow mesenchymal stem cell-derived microvesicles on glutamate injured PC12 cells]. Zhongguo Shi Yan Xue Ye Xue Za Zhi 22(4):1078–1083

Lin SS, Zhu B, Guo ZK, Huang GZ, Wang Z, Chen J, Wei XJ, Li Q (2014b) Bone marrow mesenchymal stem cell-derived microvesicles protect rat pheochromocytoma PC12 cells from glutamate-induced injury via a PI3K/Akt dependent pathway. Neurochem Res 39(5):922–931

Liu J, Kuwabara A, Kamio Y, Hu S, Park J, Hashimoto T, Lee JW (2016) Human mesenchymal stem cell-derived microvesicles prevent the rupture of intracranial aneurysm in part by suppression of mast cell activation via a PGE2-dependent mechanism. Stem Cells (Dayton, Ohio) 34(12):2943–2955

Lopez-Verrilli MA, Caviedes A, Cabrera A, Sandoval S, Wyneken U, Khoury M (2016) Mesenchymal stem cell-derived exosomes from different sources selectively promote neuritic outgrowth. Neuroscience 320:129–139

Lunn JS, Sakowski SA, Hur J, Feldman EL (2011) Stem cell technology for neurodegenerative diseases. Ann Neurol 70(3):353–361

Mokarizadeh A, Delirezh N, Morshedi A, Mosayebi G, Farshid AA, Mardani K (2012) Microvesicles derived from mesenchymal stem cells: potent organelles for induction of tolerogenic signaling. Immunol Lett 147(1–2):47–54

Mokarizadeh A, Rezvanfar MA, Dorostkar K, Abdollahi M (2013) Mesenchymal stem cell derived microvesicles: trophic shuttles for enhancement of sperm quality parameters. Reprod Toxicol (Elmsford, NY) 42:78–84

Monsel A, Zhu YG, Gennai S, Hao Q, Hu S, Rouby JJ, Rosenzwajg M, Matthay MA, Lee JW (2015) Therapeutic effects of human mesenchymal stem cell-derived microvesicles in severe pneumonia in mice. Am J Respir Crit Care Med 192(3):324–336

Monsel A, Zhu YG, Gudapati V, Lim H, Lee JW (2016) Mesenchymal stem cell derived secretome and extracellular vesicles for acute lung injury and other inflammatory lung diseases. Expert Opin Biol Ther 16(7):859–871

Moore C, Kosgodage U, Lange S, Inal JM (2017) The emerging role of exosome and microvesicle-(EMV-) based cancer therapeutics and immunotherapy. Int J Cancer 141(3):428–436

Muraro PA, Martin R, Mancardi GL, Nicholas R, Sormani MP, Saccardi R (2017) Autologous haematopoietic stem cell transplantation for treatment of multiple sclerosis. Nat Rev Neurol 13(7):391–405

Nargesi AA, Lerman LO, Eirin A (2017) Mesenchymal stem cell-derived extracellular vesicles for renal repair. Curr Gene Ther 17(1):29–42

Outka G (2009) The ethics of embryonic stem cell research and the principle of "nothing is lost". Yale J Health Policy Law Ethics 9(Suppl):585–602

Penna V, Lipay MV, Duailibi MT, Duailibi SE (2015) The likely role of proteolytic enzymes in unwanted differentiation of stem cells in culture. Future Sci OA 1(3):Fso28

Podbielska M, Banik NL, Kurowska E, Hogan EL (2013) Myelin recovery in multiple sclerosis: the challenge of remyelination. Brain Sci 3(3):1282–1324

Rahim F, Allahmoradi H, Salari F, Shahjahani M, Fard AD, Hosseini SA, Mousakhani H (2013) Evaluation of signaling pathways involved in gamma-globin gene induction using fetal hemoglobin inducer drugs. Int J Hematol Oncol Stem Cell Res 7(3):41–46

Raisi A, Azizi S, Delirezh N, Heshmatian B, Farshid AA, Amini K (2014) The mesenchymal stem cell-derived microvesicles enhance sciatic nerve regeneration in rat: a novel approach in peripheral nerve cell therapy. J Trauma Acute Care Surg 76(4):991–997

Ratajczak J, Miekus K, Kucia M, Zhang J, Reca R, Dvorak P, Ratajczak MZ (2006) Embryonic stem cell-derived microvesicles reprogram hematopoietic progenitors: evidence for horizontal transfer of mRNA and protein delivery. Leukemia 20(5):847–856

Razmkhah F, Soleimani M, Mehrabani D, Karimi MH, Kafi-Abad SA (2015) Leukemia cell microvesicles promote survival in umbilical cord blood hematopoietic stem cells. EXCLI J 14:423–429

Razmkhah F, Soleimani M, Mehrabani D, Karimi MH, Amini Kafi Abad S, Ramzi M, Iravani Saadi M, Kakoui J (2017) Leukemia microvesicles affect healthy hematopoietic stem cells. Tumour Biol 39(2):1010428317692234

Riazifar M, Pone EJ, Lotvall J, Zhao W (2017) Stem cell extracellular vesicles: extended messages of regeneration. Annu Rev Pharmacol Toxicol 57:125–154

Saki N, Jalalifar MA, Soleimani M, Hajizamani S, Rahim F (2013) Adverse effect of high glucose concentration on stem cell therapy. Int J Hematol Oncol Stem Cell Res 7(3):34–40

Sargent A, Bai L, Shano G, Karl M, Garrison E, Ranasinghe L, Planchon SM, Cohen J, Miller RH (2017) CNS disease diminishes the therapeutic functionality of bone marrow mesenchymal stem cells. Exp Neurol 295:222–232

Shahrabi S, Azizidoost S, Shahjahani M, Rahim F, Ahmadzadeh A, Saki N (2014) New insights in cellular and molecular aspects of BM niche in chronic myelogenous leukemia. Tumour Biol 35(11):10627–10633

Sideri A, Neokleous N, Brunet De La Grange P, Guerton B, Le Bousse Kerdilles MC, Uzan G, Peste-Tsilimidos C, Gluckman E (2011) An overview of the progress on double umbilical cord blood transplantation. Haematologica 96(8):1213–1220

Tian S, Liu Q, Gnatovskiy L, Ma PX, Wang Z (2015) Heart regeneration with embryonic cardiac progenitor cells and cardiac tissue engineering. J Stem Cell Transplant Biol 1(1):104

Tweedell KS (2017) The adaptability of somatic stem cells: a review. J Stem Cells Regen Med 13(1):3–13

Wang Y, Fu B, Sun X, Li D, Huang Q, Zhao W, Chen X (2015) Differentially expressed microRNAs in bone marrow mesenchymal stem cell-derived microvesicles in young and older rats and their effect on tumor growth factor-beta1-mediated epithelial-mesenchymal transition in HK2 cells. Stem Cell Res Ther 6:185

Wankhade UD, Shen M, Kolhe R, Fulzele S (2016) Advances in adipose-derived stem cells isolation, characterization, and application in regenerative tissue engineering. Stem Cells Int 2016:3206807

Wen S, Dooner M, Cheng Y, Papa E, Del Tatto M, Pereira M, Deng Y, Goldberg L, Aliotta J, Chatterjee D et al (2016) Mesenchymal stromal cell-derived extracellular vesicles rescue radiation damage to murine marrow hematopoietic cells. Leukemia 30(11):2221–2231

Xie H, Wang Z, Zhang L, Lei Q, Zhao A, Wang H, Li Q, Chen Z, Zhang W (2016a) Development of an angiogenesis-promoting microvesicle-alginate-polycaprolactone composite graft for bone tissue engineering applications. PeerJ 4:e2040

Xie H, Sun L, Zhang L, Liu T, Chen L, Zhao A, Lei Q, Gao F, Zou P, Li Q et al (2016b) Mesenchymal stem cell-derived microvesicles support ex vivo expansion of cord blood-derived CD34(+) Cells. Stem Cells Int 2016:6493241

Xie L, Mao M, Zhou L, Jiang B (2016c) Spheroid mesenchymal stem cells and mesenchymal stem cell-derived microvesicles: two potential therapeutic strategies. Stem Cells Dev 25(3):203–213

Yang B, Qiu Y, Zhou N, Ouyang H, Ding J, Cheng B, Sun J (2017) Application of stem cells in oral disease therapy: progresses and perspectives. Front Physiol 8:197

Yin H, Jiang H (2015) [Application prospect of stem cell-derived microvesicles in regeneration of injured tissues]. Sheng Wu Yi Xue Gong Cheng Xue Za Zhi 32(3):688–692

Yin G, Hu G, Wan R, Yu G, Cang X, Xiong J, Ni J, Hu Y, Xing M, Fan Y et al (2016) Role of microvesicles from bone marrow mesenchymal stem cells in acute pancreatitis. Pancreas 45(9):1282–1293

Yu B, Zhang X, Li X (2014) Exosomes derived from mesenchymal stem cells. Int J Mol Sci 15(3):4142–4157

Zhu X, You Y, Li Q, Zeng C, Fu F, Guo A, Zhang H, Zou P, Zhong Z, Wang H et al (2014) BCR-ABL1-positive microvesicles transform normal hematopoietic transplants through genomic instability: implications for donor cell leukemia. Leukemia 28(8):1666–1675

Chapter 2
Secretome: Pharmaceuticals for Cell-Free Regenerative Therapy

Nazmul Haque, Basri Johan Jeet Abdullah, and Noor Hayaty Abu Kasim

Abbreviations

AD-MSC	Adipose tissue-derived MSCs
AFSCs	Amniotic fluid stem cells
ALT	Alanine aminotransferase
AMI	Acute myocardial infarction
AM-MSCs	Amniotic membrane-derived MSCs
ANGPTs	Angiopoietins
AP-MSCs	Apical papilla-derived MSCs
Apo-PBMC	Apoptotic PBMC
AST	Aspartate aminotransferase
BDNF	Brain-derived neurotrophic factor
BMC	Bone marrow cells
BM-MSCs	MSCs from bone-marrow
BMP4	Bone morphogenetic protein 4
CNS	Central nervous system
CREB	cAMP response element-binding protein
DPSCs	Dental pulp-derived MSCs

N. Haque
Department of Restorative Dentistry, Faculty of Dentistry, University of Malaya, Kuala Lumpur, Malaysia

Department of Oral Biology and Biomedical Sciences, Faculty of Dentistry, MAHSA University, Selangor, Malaysia

B. J. J. Abdullah
Department of Biomedical Imaging, Faculty of Medicine, University of Malaya, Kuala Lumpur, Malaysia

N. H. A. Kasim (✉)
Department of Restorative Dentistry, Faculty of Dentistry, University of Malaya, Kuala Lumpur, Malaysia
e-mail: nhayaty@um.edu.my

P. V. Pham (ed.), *Stem Cell Drugs - A New Generation of Biopharmaceuticals*, Stem Cells in Clinical Applications, https://doi.org/10.1007/978-3-319-99328-7_2

EGF	Epidermal growth factor
eNOS	Endothelial nitric oxide synthase
Erk1/2	Extracellular-signal regulated kinase
ESC-MSCs	ESC-derived MSCs
ESCs	Embryonic stem cells
FB	Human fibroblasts
FGF	Fibroblast growth factor
G-CSF	Granulocyte colony stimulating factor
GDN	Glia-derived nexin
GDNF	Glial cell line-derived neurotrophic factor
GM-CSF	Granulocyte-macrophage colony stimulating factor
HDF	Human dermal fibroblast
HGF	Hepatocyte growth factor
HIF-1a	Hypoxia-inducible factor 1-alpha
HSP27	Heat shock protein 27
HUCPVC-MSCs	Human umbilical cord perivascular cell-derived MSCs
HUVECs	Human umbilical vein epithelial cells
IFN-γ	Interferon-gamma
IGF-1	Insulin-like growth factor 1
IGFBP2	Insulin-like growth factor binding protein 2
IL	Interleukin
iNOS	Inducible nitric oxide synthase
KC	Keratinocytes
KGF	Keratinocyte growth factor
LIF	Leukemia inhibitory factor
LPS	Lipopolysaccharides
MCP-1	Monocyte chemoattractant protein 1
M-CSF	Macrophage colony stimulating factor
MSCs	Mesenchymal stem cells
OM-MSCs	Olfactory mucosal MSCs
PBL	Peripheral blood leukocytes
PBMC	Peripheral blood mononuclear cells
PCNA	Proliferating cell nuclear antigen
PDGF-BB	Platelet-derived growth factor beta
PEDF	Pigment epithelium-derived factor
SCF	Stem cell factor
SDF-1	Stromal cell-derived factor-1
SM-MSCs	Skeletal muscle MSCs
sTNFR-1	Soluble TNF receptor 1
TGFβ	Transforming growth factor β
TNF-α	Tumor necrosis factor alpha
UT-MSCs	Uterine tubes MSCs
VEGF-A	Vascular endothelial growth factor A

Introduction

Over the last few decades, with the increase in life expectancy, noncommunicable and degenerative diseases such as acute myocardial infarction, stroke, diabetes, spinal cord injuries, Alzheimer's disease, and Parkinson's disease are becoming more prevalent worldwide (Christensen et al. 2009; Howse 2006). These diseases are not only considered as the top ranked causes of death but also as the major causes of morbidity that are affecting the socioeconomic and personal life of the survivors (Christensen et al. 2009; Howse 2006).

In recent years, regenerative therapy has been given considerable attention in addressing the unmet needs of treating degenerative diseases through conventional medicine. Among the different tools of regenerative medicine, embryonic stem cells (ESCs) is considered to be the best source of stem cells because of their pluripotency. However, ethical controversies over the use of ESCs, restrict their use in regenerative medicine (King and Perrin 2014; Lo and Parham 2009). Meanwhile, mesenchymal stem cells (MSCs) have shown tremendous regenerative potential and are considered as a promising tool of regenerative therapy because of their self-renewal capability and multi-differentiation potential (Estrada et al. 2013; Haque et al. 2015). Notably, several studies have shown the regenerative outcomes of MSCs based therapy despite low engraftment of the transplanted cells (Beegle et al. 2015, 2016; Malliaras and Marban 2011). This led researchers to explore the molecular mechanism behind the regenerative benefits of MSCs based therapy.

Stem cells are found to secrete a large number of paracrine factors that have mitogenic, angiogenic, antiapoptotic, antiscarring, and chemoattractant characteristics (Bollini et al. 2013; Stoddart et al. 2015). These molecules are recognized to be the possible cause behind the successful outcomes of regenerative therapy (Bollini et al. 2013; Czekanska et al. 2014; Stoddart et al. 2015). The growing evidence on the role of paracrine factors in the regeneration of affected organs has led to the introduction of cell culture supernatants or secretomes as a novel therapeutic tool of regenerative medicine.

Proteins secreted by cell, tissue, or organism under certain condition or at a particular time is expressed as "secretome" (Hathout 2007). Paracrine factors present in the secretomes help to inhibit apoptosis of cells in the damaged organs, induce proliferation of progenitor or stem cells, and induce neovascularization to supply nutrient to the affected tissues (Hathout 2007; Ratajczak et al. 2012). The role of individual or groups of paracrine factors in regeneration and regulation of various signaling pathways were being studied in the last few decades. In recent years, the regenerative potential of secretomes from stem, progenitor, and terminally differentiated cells are being studied (Haque et al. 2017; Madrigal et al. 2014; Pires et al. 2014). This is a very fast-growing field of research where the potential of secretome in all aspects of regenerative therapy are being explored in general. Hence in this chapter, the current scenario in the field of secretome research for the treatment of specific disease(s) or organ(s) will be first discussed followed by the introduction of the concept of using specific secretome composition for targeted regenerative therapy.

Sources of Secretomes

Secretome can be prepared from any cell types. To date, production of secretomes from ESCs, MSCs and other adult stem cells have been reported (Kang et al. 2009; Madrigal et al. 2014; Pires et al. 2014). Among the different types of cells used in the production of secretome, MSCs is studied most because of their immunomodulatory, multidifferentiation, and vasculogenesis potential, and trophic activity (Caplan 2013; Haque et al. 2015). More specifically, MSCs from bone-marrow (BM-MSCs), adipose tissue (AD-MSC), dental pulp (DPSCs), apical papilla (AP-MSCs), human umbilical cord perivascular cells (HUCPVC-MSCs), olfactory mucosa (OM-MSCs), skeletal muscle (SM-MSCs), uterine tubes (UT-MSCs), amniotic membrane (AM-MSCs), and ESCs (ESC-MSCs) have been used to produced secretomes in order to study their regenerative potential (Ahmed et al. 2016; Assoni et al. 2017; Bakopoulou et al. 2015; Ge et al. 2016; Lee et al. 2016; Lotfinia et al. 2016; Marfia et al. 2016; Miranda et al. 2015; Oskowitz et al. 2011; Paquet et al. 2015; Pianta et al. 2015; Pires et al. 2014; Ribeiro et al. 2011; Rossi et al. 2012; Sart et al. 2014; Teixeira et al. 2015, 2017). In addition, secretomes from amniotic fluid stem cells (hAFSCs) (Maraldi et al. 2015; Mirabella et al. 2012), peripheral blood mononuclear cells (PBMC) (Haque et al. 2017; Hoetzenecker et al. 2013; Mildner et al. 2013), apoptotic PBMC (Apo-PBMC) (Altmann et al. 2014; Hoetzenecker et al. 2012; Lichtenauer et al. 2011), monocytes (Bouchentouf et al. 2010), bone marrow cells (BMC), peripheral blood leukocytes (PBL) (Korf-Klingebiel et al. 2008), visceral endoderm like cell lines HepG2 and END2 cell line (Kang et al. 2009) have also been studied.

Regenerative Potential of Secretomes

Neuroprotection and Neurodegeneration

The term 'neurodegenerative diseases' covers both acute and chronic neurodegeneration related diseases. Damage and death of the neurons by stroke and trauma resulted in acute neurodegeneration, while chronic neurodegeneration (Alzheimer's disease, Huntington's disease, and Parkinson disease) is age related and develop gradually (Lindvall and Kokaia 2010). Both acute and chronic neuronal disorders cause functional impairment of neurons that lead to physical inability and death. Moreover, these diseases added to the social and economic burden of the patients since they need long-term care and nursing.

Regeneration of neurons in the affected part of the nervous system using secretome could be considered as a tool to treat neurodegenerative diseases (Lindvall and Kokaia 2010). In an in vitro study, activation of signaling cascades such as cAMP response element-binding protein (CREB), Akt, extracellular-signal regulated kinase (Erk1/2), and heat shock protein 27 (HSP27) that involved in the regulation of cytoprotective gene products have been detected in astrocytes and Schwann cells treated with Apo-PBMC secretome (Altmann et al. 2014). Enhanced sprouting of human primary neurons in the presence of Apo-PBMC secretome has also been reported (Altmann et al. 2014). In vivo regenerative potential of Apo-PBMC secretome using middle cerebral artery occlusion model in rat showed 37% reduction of ischemic lesion (Altmann et al. 2014). Neurotropic factors composition analysis of Apo-PBMC secretome showed significantly higher expression of brain-derived neurotrophic factor (BDNF) and this factor has been recognized to contribute toward neuronal development and function in several studies (Lu et al. 2013; Monteggia et al. 2004; Salgado et al. 2015).

Several in vitro and in vivo studies have also shown that secretomes from human MSCs possessed the potential to be neuroprotective and neuroregenerative (Ahmed et al. 2016; Assoni et al. 2017; Ge et al. 2016; Marfia et al. 2016; Pires et al. 2014; Ribeiro et al. 2011). Ahamed et al. (2016) reported markedly higher expression of vascular endothelial growth factor (VEGF), Fractalkine, RANTES, monocyte chemoattractant protein 1 (MCP-1), granulocyte-macrophage colony stimulating factor (GM-CSF), and neprilysin in the secretome from DPSCs compared to those from BM-MSCs and AD-MSCs. Decreased cytotoxicity of amyloid beta peptide to SH-SY5Y cells, and increased expression of endogenous survival factor Bcl-2 and decreased expression of apoptotic regulator Bax in SH-SY5Y cells were exhibited in the presence of secretome from DPSCs as well (Ahmed et al. 2016). Increased survival and differentiation of SH-SY5Y cells toward a neuronal phenotype have been reported in the presence of secretomes from BM-MSCs and HUCPVC-MSCs (Pires et al. 2014). Furthermore, in the presence of HUCPVC-MSCs secretome increased neuronal differentiation of human telencephalon neural precursor cells was observed (Teixeira et al. 2015). Secretome from BM-MSC was also found to support higher survival of astrocytes, microglial cells and oligodendrocytes (Ribeiro et al. 2011). However, secretomes collected at 24 and 48 h support higher survival of astrocytes and microglial cells, while secretomes collected at later time point support higher survival of oligodendrocytes (Ribeiro et al. 2011).

In an in vivo study, partial reversion of the motor phenotype and the neuronal structure in 6-hydroxidopamine induced Parkinson's disease rat was observed when treated with BM-MSC secretome (Teixeira et al. 2017). From the proteomic analysis, presence of neuroregulatory molecules, namely cystatin C, glia-derived nexin, galectin-1, pigment epithelium-derived factor (PEDF), VEGF, BDNF, interleukin-6 (IL-6), and glial cell line-derived neurotrophic factor (GDNF) were detected, hence defining its neuroregenerative potential (Teixeira et al. 2017).

Secretome from ADSC was found to inhibit the lipopolysaccharides (LPS) induced effects on microglia activation which is involved in the pathogenesis of central nervous system (CNS) inflammation (Marfia et al. 2016). Ge et al. (2016) predicted that proteins in OM-MSC secretome have neurotrophy, angiogenesis, cell growth, differentiation, apoptosis, and inflammation regulatory potential which are highly correlated with the repair of central nervous system. In addition, higher regenerative potential has been reported when secretome were used in combination with preconditioned stem cells. Sart et al. (2014) have shown that preconditioning of ESC-derived neural progenitor cells aggregates in hypoxic environment in the presence of BM-MSC secretome enhances the engraftment potential and neurogenesis of cells following transplantation (Sart et al. 2014). A cocktail of secretomes has also been studied in vitro, where pooled secretomes from AD-MSCs, SM-MSCs, and UT-MSCs from five different donors was shown to delay apoptosis and enhance migration of Duchenne muscular dystrophy myoblasts (Assoni et al. 2017).

Angiogenesis

Angiogenesis is vital in repair and regeneration of affected tissues or organs, and tissue engineering. Identification of angiogenic factors and their presence in the secretomes from different cell sources has been reported (Bakopoulou and About 2016; Burrows et al. 2013; Konala et al. 2016; Newman et al. 2013). An ex vivo study demonstrated longer neovascular sprouts generation from rat aortic rings cultured in serum deprived BM-MSC secretome compared to the control group. In vitro angiogenesis assay also showed the superiority of serum deprived BM-MSC secretome. The authors attributed the results to the higher expression of VEGF-A, angiopoietins (ANGPTs), insulin-like growth factor 1 (IGF-1), and hepatocyte growth factor (HGF) in the BM-MSC secretome yielded from serum deprived culture condition (Oskowitz et al. 2011). Similarly, significantly higher expression of angiogenic mediators (VEGF-A, VEGF-C, IL-8, RANTES, and MCP-1) and lower expression of immunomodulatory mediators (IL-1b, IL-6, IL-1Ra, IL-15, and FGF-2 and HGF) was observed in the secretome from BM-MSCs cultured in anoxic (0.1% oxygen) compared to normoxic and hypoxic (5% oxygen) conditions (Paquet et al. 2015). Both in vitro and in vivo studies also showed significantly better chemoattractant and angiogenic potential of the BM-MSC secretome derived from anoxic condition (Paquet et al. 2015). In another study, AP-MSCs were cultured in serum-deprived, glucose deprived, and hypoxic condition individually or in combination. Finally, it was found that higher numbers and amounts of proangiogenic (angiogenin, IGFBP-3, VEGF) and lower amounts of antiangiogenic factors (serpin-E1, TIMP-1, TSP-1) were secreted when cultured in all stressed conditions combined compared to partial combinations or in one stressed condition only (Bakopoulou et al. 2015). Furthermore, the secretome obtained was most effective in supporting migration and formation of capillary like structure by human umbilical vein epithelial cells (HUVECs) (Bakopoulou et al. 2015). These results substantiate the necessity of utilizing preconditioning strategies to enhance the angiogenic potential of secretomes produced from MSCs regardless of their sources.

Cardiac Regeneration and Cardio-Protection

Both human and murine monocytes cultured in angiogenic conditions were found to express significantly higher amount of HGF, IGF-1, MCP-1, and soluble TNF receptor 1 (sTNFR-1) compared to their precursors (Bouchentouf et al. 2010).

They also demonstrated the presence of HGF, IGF-1, and sTNFR-1 in the secretome yielded from monocytes cultured in angiogenic condition, and the secretome reduces tumor necrosis factor alpha (TNF-α), staurosporine, and oxidative stress induced death of murine HL-1 cardiomyocyte cell line. However, the presence of HGF, IGF-1, and MCP-1 in this secretome helped to promote endothelial cell proliferation and capacity to form vessels those are needed for cardiac remodeling (Bouchentouf et al. 2010).

Secretome from Apo-PBMC was found to reduce microvascular obstruction during acute myocardial infarction (AMI) in pigs and the platelet activation markers was also lowered in the plasma sample collected (Hoetzenecker et al. 2012). They further confirmed their findings using an in vitro study, where Apo-PBMC secretome caused impaired activation and aggregation of human and pig platelets. In addition, increased vasodilation capacity via activation of endothelial nitric oxide synthase (eNOS) and inducible nitric oxide synthase (iNOS) was also reported in the presence of secretome from Apo-PBMC (Hoetzenecker et al. 2012).

In another in vitro study, induction of caspase-8-dependent apoptosis in autoreactive $CD4^+$ T cell in the presence of PBMC secretome was observed. This result supports the notion that secretome from PBMC could potentially be used for treatment of inflammatory heart diseases (Hoetzenecker et al. 2013).

Secretome from Apo-PBMC have also been shown to exhibit cardioprotective effect through a combination of in vivo and in vitro studies. In experimental AMI rat and pig models, secretome from Apo-PBMC reduced scar tissue formation (Lichtenauer et al. 2011). While in porcine closed chest reperfused AMI model, higher values of ejection fraction, a better cardiac output and a reduced extent of infarct size were reported. Induced activation of prosurvival signaling-cascade (AKT, Erk1/2, CREB, c-jun), increased antiapoptotic gene products (Bcl-2, BAGI) and reduced starvation-induced cell death was seen in human cardiomyocytes in the presence of the Apo-PBMC secretome in vitro (Lichtenauer et al. 2011).

Secretomes from BMC and PBL both have shown stimulated human coronary artery endothelial cell proliferation, migration, and tube formation, and induced cell sprouting in mouse aortic ring assay (Korf-Klingebiel et al. 2008). Both secretomes were also found to protect rat ventricular cardiomyocytes from cell death induced by simulated ischemia or ischemia followed by reperfusion. Notably, a combination of the BMC and PBL secretomes showed a synergistic effect (Korf-Klingebiel et al. 2008).

Acute Liver Failure

Recently, Lotfinia et al. (2016) studied the potential of the secretomes from ESC-MSC and BM-MSC for the treatment of inflammatory hepatic conditions (Lotfinia et al. 2016). In their study, significantly upregulated expression of angiogenin, IGFBP2, transforming growth factor β1 (TGFβ1), and MCP1 was observed in the

ESC-MSC secretome compared to that in BM-MSC secretome. However, among the 174 proteins analyzed, most of the cytokines in BM-MSC secretome showed higher expression than ESC-MSC secretome. VEGF and bone morphogenetic protein 4 (BMP4) which are involved in the regulation of immune regulation, epithelial cell proliferation, and negative regulation of apoptosis were expressed in the both secretomes. Compared to the control group, both secretomes were found to increase in vitro viability of hepatocytes, and decrease aspartate aminotransferase (AST) and alanine aminotransferase (ALT) in the serum from the thioacetamide-induced acute liver failure mice. In addition, immunomodulatory potential of ESC-MSC secretome was better than BM-MSC secretome as indicated by the increased IL-10 secretion. However, none of the secretome showed any effect on the survival of acute liver failure induced mice after 1 week (Lotfinia et al. 2016).

AD-MSC secretome obtained from hypoxic culture conditions showed significantly higher expression of hypoxia-inducible factor 1-alpha (HIF-1α), HGF, and VEGF compared to those collected at normoxic condition (Lee et al. 2016). AD-MSC secretome collected at hypoxic condition increased proliferating cell nuclear antigen (PCNA) marker expression and proliferation of AML12 cells. While, decreased level of IL-6, TNF-α, AST, and ALT in the serum of partially hepatectomized mice, and increased PCNA expression and the number of KI-67 positive cells in the hepatectomized liver was also reported (Lee et al. 2016).

Osteogenic and Chondrogenic Differentiation

Secretomes from visceral endoderm like cell lines HepG2 and END2 cell line have shown osteogenic and chondrogenic differentiation potential. Presence of six common protein (β-actin, complement component 3, fibronectin1, immunoglobulin, vimentin, and vinculin) required for the migration and adhesion of cells was detected in the both secretomes (Kang et al. 2009). Though there are lack of studies on the osteogenic and chondrogenic regeneration using secretomes; role of different paracrine factors, namely TGF-β, stromal cell-derived factor-1 (SDF-1), HGF, fibroblast growth factor (FGF) 18, and IGF-1 in osteogenesis and chondrogenesis have been acknowledged by several researchers (Correa et al. 2015; Jenniskens et al. 2006; Stoddart et al. 2015; Takebayashi et al. 1995).

Immunoregulation

Immunosuppression or immunoregulation is highly needed to control autoimmune diseases or prevent rejection of allogenic implants. Secretome from AM-MSCs was found to modulate lymphocyte proliferation in a dose-dependent manner

(Rossi et al. 2012). Further studies confirmed that secretome from AM-MSCs suppressed the proliferation of both $CD4^+$ T-helper (Th) and $CD8^+$ cytotoxic T-lymphocytes, and also showed inhibitory properties on both central and effector memory subsets (Pianta et al. 2015). More specifically, AM-MSC secretome significantly reduced the expression of markers associated to the Th1 and Th17 populations, while no effect on the Th2 population was reported. Notably, AM-MSC secretome significantly induced Treg cells, and it was further confirmed by the increased secretion of TGF-β (Pianta et al. 2015). Immunomodulatory potential of secretome from AFSCs has also been reported (Maraldi et al. 2015).

Wound Healing

Secretome has also been shown to have wound healing potential. Miranda et al. (2015) reported that secretomes from both UC-MSCs and BM-MSCs have an effect on the migration of human dermal fibroblast (HDF) and keratinocyte (HaCaT). However, secretome from UC-MSCs showed significantly higher migration of HaCaTs compared to HDFs, while the opposite effect was observed in the secretome from BM-MSCs (Miranda et al. 2015). The migration of keratinocytes in the presence of UC-MSC secretome were linked to the relatively higher presence of epidermal growth factor (EGF), FGF-2, and keratinocyte growth factor (KGF). This study showed the potential of UC-MSC secretome in maintaining the earlier homeostasis and inflammation stages of wound healing, while the BM-MSC secretome could be useful in promoting later proliferative and final remodeling of tissues that is linked to the presence of granulocyte colony stimulating factor (G-CSF), IL-6, VEGF-A, TGF-β in it (Miranda et al. 2015).

Mirabella et al. (2012) also elucidated the wound healing potential of secretome from AFSCs through an in vivo study. In their study, raised flaps treated with AFSCs secretome showed 50% higher perfusion on day 7 post-operation than the baseline, and subsequently necrosis development was delayed. Moreover, normal arrangement of epidermal and dermal structures and a high density of vessels in subcutaneous tissues were observed histologically (Mirabella et al. 2012). AFSCs secretome also induces the migration of wound and scar repairing $CD31^+$/$VEGFR2^+$ and $CD31^+$/$CD34^+$ cells into the ischemic subcutaneous tissues (Mirabella et al. 2012).

Significantly rapid wound closure and reepithelialization was observed in the skin of full-thickness punch biopsy wound modeled rat when treated with PBMC secretome containing emulsion. Meanwhile, increased CD31 positive cell population indicated enhanced neoangiogenesis at the site of PBMC secretome treated tissue (Mirabella et al. 2012). PBMC secretome also induced migration of primary human fibroblasts (FB) and keratinocytes (KC) in vitro. However, no effect on the proliferation of these cell populations was seen. Notably, induced proliferation and angiogenic tube formation of endothelial cells in the presence of PBMC secretome was also reported. These result supports the potential use of PBMC secretome in treating non-healing skin ulcers (Mildner et al. 2013).

Secretome as Cell-Free Pharmaceuticals for Tissue-Specific Regeneration

The discussion in the earlier sections indicates that the regenerative potential of secretomes from different cell sources is highly dependent on the paracrine factors present. Biological functions of some common regenerative paracrine factors are listed in Table 2.1.

Table 2.1 Major biological functions of some selected paracrine factors

Name of the paracrine factors	Function (References)
Brain-derived neurotrophic factor (BDNF)	• Promotes survival of neurons, synaptogenesis, and synaptic plasticity (Lu et al. 2013).
Epidermal growth factor (EGF)	• Regulates cellular proliferation, differentiation, survival, and motility (Herbst 2004). • Regulates proliferation of MSCs isolated from different origins while maintaining their regenerative potential (Hu et al. 2013; Tamama et al. 2006, 2010).
Fibroblast growth factor 2 (FGF-2)	• Promotes angiogenesis, survival of cells, and wound healing (Beenken and Mohammadi 2009). • Stimulates migration and proliferation of endothelial cells (Beenken and Mohammadi 2009). • Encourages mitogenesis of smooth muscle cells and fibroblasts (Beenken and Mohammadi 2009). • Shows a broad spectrum of mitogenic effects (Salcedo et al. 1999; Werner and Grose 2003). • Stimulates the in vitro expansion of human BM-MSCs by activation of JNK signaling (Ahn et al. 2009). • Slows down the ageing process of MSCs by decreasing the gradual loss of telomere sequences (Bianchi et al. 2003; Yanada et al. 2006). • Cytoprotective role of FGFs have also been acknowledged by researchers (Werner and Grose 2003). • Increases expression of CXCR4 on human endothelial cells and help in angiogenesis (Salcedo et al. 1999).
Granulocyte colony stimulating factor (G-CSF)	• Regulates granulopoiesis (Zhang et al. 2009). • Promotes survival, proliferation, activation, and maturation of hematopoietic progenitors of neutrophil lineage (Zhang et al. 2009). • Promotes cellular proliferation and migration, and prevents apoptosis (Murakami et al. 2013). • Mobilizes HSC and MSCs from bone marrow (Kawada et al. 2004). • Improves chemotactic property of MSCs in vitro (Murakami et al. 2013).

(continued)

Table 2.1 (continued)

Name of the paracrine factors	Function (References)
Granulocyte-macrophage colony stimulating factor (GM-CSF)	• Stimulates proliferation and differentiation of hematopoietic progenitors (Shi et al. 2006). • Acts as chemoattractant and induces mobilization of progenitors in the circulation (Rojas et al. 2005).
Hepatocyte growth factor (HGF)	• Mitogenic for epithelial and endothelial cells (Sulpice et al. 2009). • Promotes angiogenesis; induces kidney and liver regeneration (Galimi et al. 2001; Sulpice et al. 2009). • Promotes proliferation and survival of various cell types (Forte et al. 2006). • Induces migration and site-specific homing of various cell types including MSCs from different origins (Son et al. 2006; Sulpice et al. 2009). • Helps in immunomodulation (Maraldi et al. 2015).
Leukemia inhibitory factor (LIF)	• Inhibits proliferation and induces differentiation of macrophages (Moon et al. 2002). • Promotes neuronal survival and differentiation (Moon et al. 2002). • Stimulates glial development (Moon et al. 2002). • Helps to maintain self-renewal and multidifferentiation potential of various stem cells including MSCs (Kolf et al. 2007; Metcalf 2003).
Macrophage colony stimulating factor (M-CSF)	• Regulates production, survival, and function of monocytes, macrophages, and osteoclasts (Grasset et al. 2010).
Platelet-derived growth factor beta (PDGF-BB)	• Induces fibroblast proliferation, collagen production, and angiogenesis (Andrae et al. 2008). • Promotes wound healing (Andrae et al. 2008). • Influences periodontal regeneration (Andrae et al. 2008). • Induces both expansion and migration of MSCs (Fierro et al. 2007; Tamama et al. 2006). • Helps survival of MSCs as well (Krausgrill et al. 2009).
Stem cell factor (SCF), KIT ligand	• Promotes survival, proliferation, and differentiation of hematopoietic stem cells and progenitor cells (Broudy 1997). • Promote survival of mature cells as well (Broudy 1997). • Regulates the migration, differentiation, and proliferation of several cell types (Lennartsson and Rönnstrand 2012). • Induces the migration and homing of MSCs (Pan et al. 2013).
Stromal cell-derived factor-1a (SDF-1A)	• Induces migration of neutrophils to site of infection (Murphy et al. 2007). • Promotes mobilization and directed migration of stem cells (Murphy et al. 2007). • Influences neurogenesis (Murphy et al. 2007). • Helps site-specific migration and homing of MSCs and other cells through chemokine receptor CXCR4 (He et al. 2010; Yu et al. 2015).
Tumor necrosis factor alpha (TNF-α)	• Induces tumor cell apoptosis, inflammation, and immune response (Pfeffer 2003).
Vascular endothelial growth factor A (VEGF-A)	• Shows angiogenic, arteriogenic, antiapoptotic, and immunoregulatory properties (Sulpice et al. 2009; Wang et al. 2006). • Increases proliferation and survival MSCs (Pons et al. 2008).

(continued)

Table 2.1 (continued)

Name of the paracrine factors	Function (References)
Interferon-gamma (IFN-γ)	• Induces antigen processing and presentation (Schroder et al. 2004). • Inhibit proliferation and induce apoptosis (Schroder et al. 2004). • Induce immunomodulation and leukocyte trafficking (Schroder et al. 2004).
Interleukin 2 (IL-2)	• Regulates proliferation, activation, and differentiation of lymphocytes (Liao et al. 2011).
Interleukin 3 (IL-3)	• Promotes proliferation and differentiation of hematopoietic progenitors (Nitsche et al. 2003).
Interleukin 6 (IL-6)	• Promotes angiogenesis, wound healing, and cell migration (Yew et al. 2011). • Promotes axon regeneration (Leibinger et al. 2013). • Stimulates the production of acute phase proteins (Fattori et al. 1994). • Favors chronic inflammatory responses by stimulating T- and B-lymphocytes (Gabay 2006).
Interleukin 10 (IL-10)	• Inhibits Th1 cells, natural killer cells, and macrophages (Couper et al. 2008). • Enhances proliferation, survival, and antibody production of B cells (Rousset et al. 1992). • Promotes immunosuppressive functions (Pierson and Liston 2010).
Interleukin 12 (IL-12p70)	• Increases IFN-γ production (Del Vecchio et al. 2007). • Induces Th1 differentiation (Del Vecchio et al. 2007). • Promotes proliferation and cytolytic activity of natural killer and T cells (Del Vecchio et al. 2007).
Interleukin 23 (IL-23)	• Induces autoimmunity (Gaffen et al. 2014). • Induces tissue destruction (Gaffen et al. 2014).

To date, in vitro and in vivo studies conducted on the regenerative application of secretome appeared to be rather subjective and the outcomes varied. The variation in outcomes could be attributed to the donors, cell types and incubation times (Assoni et al. 2017; Haque et al. 2017). Therefore, maintaining batch to batch consistency of paracrine factors' composition in the secretome will be very challenging.

Based on the research outcomes described above, we attempted to identify and select the vital paracrine factors needed to yield the best regenerative outcome for a particular disease or organ type. Following the analysis of the paracrine factors' composition in the different secretomes regardless of their sources, we were able to group them and proposed its use for targeted regenerative therapies (Fig. 2.1). Further studies and precise grouping of the paracrine factors would be more effective in sorting and selecting a secretome-type for a targeted tissue-based regeneration and finally engineering secretome to be "cell-free pharmaceuticals" in the near future. Pretreatment of cells (Bakopoulou et al. 2015; Sart et al. 2014) and the usage of dynamic culture conditions (Teixeira et al. 2016) could even be used to regulate the production of targeted paracrine factors in the large-scale production of secretome.

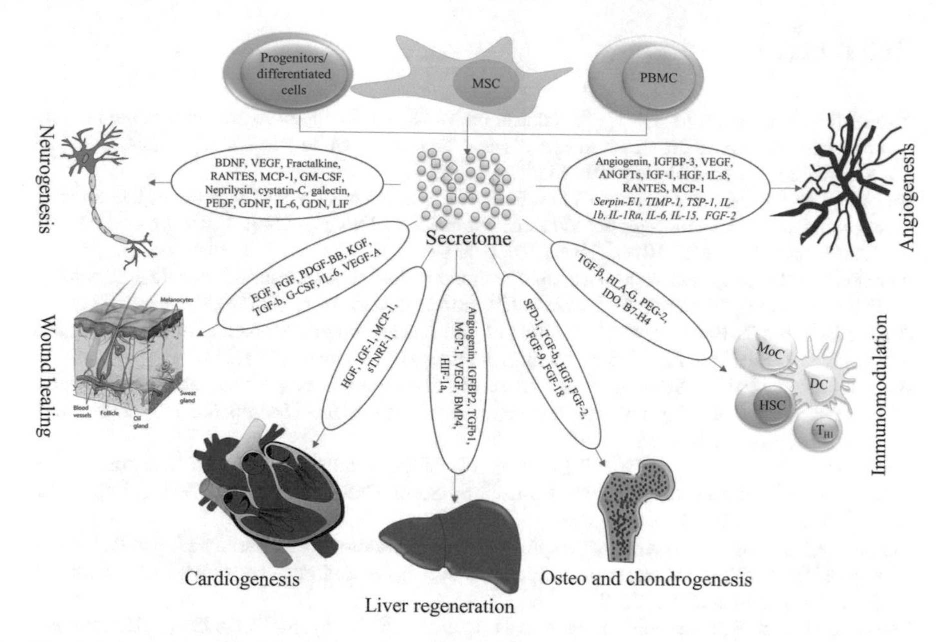

Fig. 2.1 Paracrine factors' composition proposed for targeted tissue or organ-based regenerative therapy. *ANGPTs* angiopoietins, *BDNF* brain-derived neurotrophic factor, *BMP4* bone morphogenetic protein 4, *EGF* epidermal growth factor, *FGF* fibroblast growth factor, *G-CSF* granulocyte colony stimulating factor, *GDN* glia-derived nexin, *GDNF* glial cell line-derived neurotrophic factor, *GM-CSF* granulocyte-macrophage colony stimulating factor, *HGF* hepatocyte growth factor, *HIF-1a* hypoxia-inducible factor 1-alpha, *IGF-1* insulin-like growth factor 1, *IGFBP2* insulin-like growth factor binding protein 2, *IL* interleukin, *LIF* leukemia inhibitory factor, *MCP-1* monocyte chemoattractant protein 1, *MSCs* mesenchymal stem cells, *PBMC* peripheral blood mononuclear cells, *PDGF-BB* platelet-derived growth factor beta, *PEDF* pigment epithelium-derived factor, *SDF-1* stromal cell-derived factor-1, *sTNFR-1* soluble TNF receptor 1, *TGFβ* transforming growth factor β, *VEGF-A* vascular endothelial growth factor A (Regular and italic fonts denote expected higher and lower expression of the paracrine factors in the secretome respectively.)

Conclusion

The presence of paracrine factors in the secretome plays a vital role in the process of regeneration. From the critical analysis of the outcomes based on in vitro and in vivo studies of secretome and the molecules involved in the regenerative process, we attempted to categorize the paracrine factors. Finally, we proposed that regardless of the source of the secretome and on the basis of the presence of the group of paracrine factors, secretome could be selected for targeted regenerative therapy.

Acknowledgment This work was supported by High Impact Research MOHE Grant UM.C/625/1/HIR/MOHE/DENT/01 from the Ministry of Higher Education Malaysia and University of Malaya Research Grant UMRG RP019/13HTM.

Conflicts of Interest: The authors confirm that there are no conflicts of interest related to this study.

References

Ahmed NM, Murakami M, Hirose Y, Nakashima M (2016) Therapeutic potential of dental pulp stem cell secretome for Alzheimer's disease treatment: an in vitro study. Stem Cells Int 2016:8102478. https://doi.org/10.1155/2016/8102478

Ahn H-J, Lee W-J, Kwack K, Kwon YD (2009) FGF2 stimulates the proliferation of human mesenchymal stem cells through the transient activation of JNK signaling. FEBS Lett 583:2922–2926. https://doi.org/10.1016/j.febslet.2009.07.056

Altmann P et al (2014) Secretomes of apoptotic mononuclear cells ameliorate neurological damage in rats with focal ischemia. F1000Res 3:131. https://doi.org/10.12688/f1000research.4219.2

Andrae J, Gallini R, Betsholtz C (2008) Role of platelet-derived growth factors in physiology and medicine. Genes Dev 22:1276–1312. https://doi.org/10.1101/gad.1653708

Assoni A et al (2017) Different donors mesenchymal stromal cells secretomes reveal heterogeneous profile of relevance for therapeutic use. Stem Cells Dev 26:206–214. https://doi.org/10.1089/scd.2016.0218

Bakopoulou A, About I (2016) Stem cells of dental origin: current research trends and key milestones towards clinical application. Stem Cells Int 2016:4209891. https://doi.org/10.1155/2016/4209891

Bakopoulou A et al (2015) Angiogenic potential and secretome of human apical papilla mesenchymal stem cells in various stress microenvironments. Stem Cells Dev 24:2496–2512. https://doi.org/10.1089/scd.2015.0197

Beegle J, Lakatos K, Kalomoiris S, Stewart H, Isseroff RR, Nolta JA, Fierro FA (2015) Hypoxic preconditioning of mesenchymal stromal cells induces metabolic changes, enhances survival, and promotes cell retention in vivo. Stem Cells 33:1818–1828. https://doi.org/10.1002/stem.1976

Beegle JR et al (2016) Preclinical evaluation of mesenchymal stem cells overexpressing VEGF to treat critical limb ischemia. Mol Ther Methods Clin Dev 3:16053. https://doi.org/10.1038/mtm.2016.53

Beenken A, Mohammadi M (2009) The FGF family: biology, pathophysiology and therapy. Nat Rev Drug Discov 8:235–253. https://doi.org/10.1038/nrd2792

Bianchi G, Banfi A, Mastrogiacomo M, Notaro R, Luzzatto L, Cancedda R, Quarto R (2003) Ex vivo enrichment of mesenchymal cell progenitors by fibroblast growth factor 2. Exp Cell Res 287:98–105. https://doi.org/10.1016/S0014-4827(03)00138-1

Bollini S, Gentili C, Tasso R, Cancedda R (2013) The regenerative role of the fetal and adult stem cell secretome. J Clin Med 2:302–327. https://doi.org/10.3390/jcm2040302

Bouchentouf M et al (2010) Monocyte derivatives promote angiogenesis and myocyte survival in a model of myocardial infarction. Cell Transplant 19:369–386. https://doi.org/10.3727/096368909x484266

Broudy VC (1997) Stem cell factor and hematopoiesis. Blood 90:1345–1364

Burrows GG et al (2013) Dissection of the human multipotent adult progenitor cell secretome by proteomic analysis. Stem Cells Transl Med 2:745–757. https://doi.org/10.5966/sctm.2013-0031

Caplan AI (2013) MSCs as therapeutics. In: Hematti P, Keating A (eds) Mesenchymal stromal cells. Springer, New York, pp 79–90

Christensen K, Doblhammer G, Rau R, Vaupel JW (2009) Ageing populations: the challenges ahead. Lancet 374:1196–1208. https://doi.org/10.1016/S0140-6736(09)61460-4

Correa D et al (2015) Sequential exposure to fibroblast growth factors (FGF) 2, 9 and 18 enhances hMSC chondrogenic differentiation. Osteoarthritis Cartilage 23:443–453. https://doi.org/10.1016/j.joca.2014.11.013

Couper KN, Blount DG, Riley EM (2008) IL-10: the master regulator of immunity to infection. J Immunol 180:5771–5777. https://doi.org/10.4049/jimmunol.180.9.5771

Czekanska EM, Ralphs JR, Alini M, Stoddart MJ (2014) Enhancing inflammatory and chemotactic signals to regulate bone regeneration. Eur Cell Mater 28:320–334

Del Vecchio M, Bajetta E, Canova S, Lotze MT, Wesa A, Parmiani G, Anichini A (2007) Interleukin-12: biological properties and clinical application. Clin Cancer Res 13:4677–4685. https://doi.org/10.1158/1078-0432.CCR-07-0776

Estrada JC et al (2013) Human mesenchymal stem cell-replicative senescence and oxidative stress are closely linked to aneuploidy. Cell Death Dis 4:e691. https://doi.org/10.1038/cddis.2013.211

Fattori E et al (1994) Defective inflammatory response in interleukin 6-deficient mice. J Exp Med 180:1243–1250. https://doi.org/10.1084/jem.180.4.1243

Fierro F, Illmer T, Jing D, Schleyer E, Ehninger G, Boxberger S, Bornhäuser M (2007) Inhibition of platelet-derived growth factor receptorβ by imatinib mesylate suppresses proliferation and alters differentiation of human mesenchymal stem cells in vitro. Cell Prolif 40:355–366. https://doi.org/10.1111/j.1365-2184.2007.00438.x

Forte G et al (2006) Hepatocyte growth factor effects on mesenchymal stem cells: proliferation, migration, and differentiation. Stem Cells 24:23–33. https://doi.org/10.1634/stemcells.2004-0176

Gabay C (2006) Interleukin-6 and chronic inflammation. Arthritis Res Ther 8:1–6. https://doi.org/10.1186/ar1917

Gaffen SL, Jain R, Garg AV, Cua DJ (2014) The IL-23-IL-17 immune axis: from mechanisms to therapeutic testing. Nat Rev Immunol 14:585–600. https://doi.org/10.1038/nri3707

Galimi F et al (2001) Hepatocyte growth factor is a regulator of monocyte-macrophage function. J Immunol 166:1241–1247

Ge LT et al (2016) Secretome of olfactory mucosa mesenchymal stem cell, a multiple potential stem cell. Stem Cells Int 2016:1243659. https://doi.org/10.1155/2016/1243659

Grasset MF, Gobert-Gosse S, Mouchiroud G, Bourette RP (2010) Macrophage differentiation of myeloid progenitor cells in response to M-CSF is regulated by the dual-specificity phosphatase DUSP5. J Leukoc Biol 87:127–135. https://doi.org/10.1189/jlb.0309151

Haque N, Kasim NH, Rahman MT (2015) Optimization of pre-transplantation conditions to enhance the efficacy of mesenchymal stem cells. Int J Biol Sci 11:324–334. https://doi.org/10.7150/ijbs.10567

Haque N, Kasim NHA, Kassim NLA, Rahman MT (2017) Autologous serum supplement favours in vitro regenerative paracrine factors synthesis. Cell Prolif 50. https://doi.org/10.1111/cpr.12354

Hathout Y (2007) Approaches to the study of the cell secretome. Expert Rev Proteomics 4:239–248. https://doi.org/10.1586/14789450.4.2.239

He X, Ma J, Jabbari E (2010) Migration of marrow stromal cells in response to sustained release of stromal-derived factor-1alpha from poly(lactide ethylene oxide fumarate) hydrogels. Int J Pharm 390:107–116. https://doi.org/10.1016/j.ijpharm.2009.12.063

Herbst RS (2004) Review of epidermal growth factor receptor biology. Int J Radiat Oncol Biol Phys 59:S21–S26. https://doi.org/10.1016/j.ijrobp.2003.11.041

Hoetzenecker K et al (2012) Secretome of apoptotic peripheral blood cells (APOSEC) attenuates microvascular obstruction in a porcine closed chest reperfused acute myocardial infarction model: role of platelet aggregation and vasodilation. Basic Res Cardiol 107:292. https://doi.org/10.1007/s00395-012-0292-2

Hoetzenecker K et al (2013) Mononuclear cell secretome protects from experimental autoimmune myocarditis. Eur Heart J 36:ehs459. https://doi.org/10.1093/eurheartj/ehs459

Howse K (2006) Increasing life expectancy and the compression of morbidity: a critical review of the debate. Oxford Institute of Population Ageing, Oxford

Hu F, Wang X, Liang G, Lv L, Zhu Y, Sun B, Xiao Z (2013) Effects of epidermal growth factor and basic fibroblast growth factor on the proliferation and osteogenic and neural differentiation of adipose-derived stem cells. Cell Reprogram 15:224–232. https://doi.org/10.1089/cell.2012.0077

Jenniskens YM et al (2006) Biochemical and functional modulation of the cartilage collagen network by IGF1, TGFβ2 and FGF2. Osteoarthritis Cartilage 14:1136–1146. https://doi.org/10.1016/j.joca.2006.04.002

Kang YY, Nagy JM, Polak JM, Mantalaris A (2009) Proteomic characterization of the conditioned media produced by the visceral endoderm-like cell lines HepG2 and END2: toward a defined medium for the osteogenic/chondrogenic differentiation of embryonic stem cells. Stem Cells Dev 18:77–91. https://doi.org/10.1089/scd.2008.0026

Kawada H et al (2004) Nonhematopoietic mesenchymal stem cells can be mobilized and differentiate into cardiomyocytes after myocardial infarction. Blood 104:3581–3587. https://doi.org/10.1182/blood-2004-04-1488

King NM, Perrin J (2014) Ethical issues in stem cell research and therapy. Stem Cell Res Ther 5:85. https://doi.org/10.1186/scrt474

Kolf CM, Cho E, Tuan RS (2007) Mesenchymal stromal cells. Biology of adult mesenchymal stem cells: regulation of niche, self-renewal and differentiation. Arthritis Res Ther 9:1–10. https://doi.org/10.1186/ar2116

Konala VB, Mamidi MK, Bhonde R, Das AK, Pochampally R, Pal R (2016) The current landscape of the mesenchymal stromal cell secretome: a new paradigm for cell-free regeneration. Cytotherapy 18:13–24. https://doi.org/10.1016/j.jcyt.2015.10.008

Korf-Klingebiel M et al (2008) Bone marrow cells are a rich source of growth factors and cytokines: implications for cell therapy trials after myocardial infarction. Eur Heart J 29:2851–2858. https://doi.org/10.1093/eurheartj/ehn456

Krausgrill B et al (2009) Influence of cell treatment with PDGF-BB and reperfusion on cardiac persistence of mononuclear and mesenchymal bone marrow cells after transplantation into acute myocardial infarction in rats. Cell Transplant 18:847–853. https://doi.org/10.3727/096368909X471134

Lee SC, Jeong HJ, Lee SK, Kim SJ (2016) Hypoxic conditioned medium from human adipose-derived stem cells promotes mouse liver regeneration through JAK/STAT3 signaling. Stem Cells Transl Med 5:816–825. https://doi.org/10.5966/sctm.2015-0191

Leibinger M, Muller A, Gobrecht P, Diekmann H, Andreadaki A, Fischer D (2013) Interleukin-6 contributes to CNS axon regeneration upon inflammatory stimulation. Cell Death Dis 4:e609. https://doi.org/10.1038/cddis.2013.126

Lennartsson J, Rönnstrand L (2012) Stem cell factor receptor/c-Kit: from basic science to clinical implications. Physiol Rev 92:1619–1649. https://doi.org/10.1152/physrev.00046.2011

Liao W, Lin JX, Leonard WJ (2011) IL-2 family cytokines: new insights into the complex roles of IL-2 as a broad regulator of T helper cell differentiation. Curr Opin Immunol 23:598–604. https://doi.org/10.1016/j.coi.2011.08.003

Lichtenauer M et al (2011) Secretome of apoptotic peripheral blood cells (APOSEC) confers cytoprotection to cardiomyocytes and inhibits tissue remodelling after acute myocardial infarction: a preclinical study. Basic Res Cardiol 106:1283–1297. https://doi.org/10.1007/s00395-011-0224-6

Lindvall O, Kokaia Z (2010) Stem cells in human neurodegenerative disorders — time for clinical translation? J Clin Invest 120:29–40. https://doi.org/10.1172/JCI40543

Lo B, Parham L (2009) Ethical issues in stem cell research. Endocr Rev 30:204–213. https://doi.org/10.1210/er.2008-0031

Lotfinia M et al (2016) Effect of secreted molecules of human embryonic stem cell-derived mesenchymal stem cells on acute hepatic failure model. Stem Cells Dev 25:1898–1908. https://doi.org/10.1089/scd.2016.0244

Lu B, Nagappan G, Guan X, Nathan PJ, Wren P (2013) BDNF-based synaptic repair as a disease-modifying strategy for neurodegenerative diseases. Nat Rev Neurosci 14:401–416. https://doi.org/10.1038/nrn3505

Madrigal M, Rao K, Riordan N (2014) A review of therapeutic effects of mesenchymal stem cell secretions and induction of secretory modification by different culture methods. J Transl Med 12:260. https://doi.org/10.1186/s12967-014-0260-8

Malliaras K, Marban E (2011) Cardiac cell therapy: where we've been, where we are, and where we should be headed. Br Med Bull 98:161–185. https://doi.org/10.1093/bmb/ldr018

Maraldi T, Beretti F, Guida M, Zavatti M, De Pol A (2015) Role of hepatocyte growth factor in the immunomodulation potential of amniotic fluid stem cells. Stem Cells Transl Med 4:539–547. https://doi.org/10.5966/sctm.2014-0266

Marfia G et al (2016) The adipose mesenchymal stem cell secretome inhibits inflammatory responses of microglia: evidence for an involvement of sphingosine-1-phosphate signalling. Stem Cells Dev 25:1095–1107. https://doi.org/10.1089/scd.2015.0268

Metcalf D (2003) The unsolved enigmas of leukemia inhibitory factor. Stem Cells 21:5–14. https://doi.org/10.1634/stemcells.21-1-5

Mildner M et al (2013) Secretome of peripheral blood mononuclear cells enhances wound healing. PLoS One 8:e60103. https://doi.org/10.1371/journal.pone.0060103

Mirabella T, Hartinger J, Lorandi C, Gentili C, van Griensven M, Cancedda R (2012) Proangiogenic soluble factors from amniotic fluid stem cells mediate the recruitment of endothelial progenitors in a model of ischemic fasciocutaneous flap. Stem Cells Dev 21:2179–2188. https://doi.org/10.1089/scd.2011.0639

Miranda JP et al (2015) The human umbilical cord tissue-derived msc population UCX (R) promotes early motogenic effects on keratinocytes and fibroblasts and G-CSF-mediated mobilization of BM-MSCs when transplanted in vivo. Cell Transplant 24:865–877. https://doi.org/10.3727/096368913x676231

Monteggia LM et al (2004) Essential role of brain-derived neurotrophic factor in adult hippocampal function. Proc Natl Acad Sci U S A 101:10827–10832. https://doi.org/10.1073/pnas.0402141101

Moon C, Yoo JY, Matarazzo V, Sung YK, Kim EJ, Ronnett GV (2002) Leukemia inhibitory factor inhibits neuronal terminal differentiation through STAT3 activation. Proc Natl Acad Sci U S A 99:9015–9020. https://doi.org/10.1073/pnas.132131699

Murakami M et al (2013) The use of granulocyte-colony stimulating factor induced mobilization for isolation of dental pulp stem cells with high regenerative potential. Biomaterials 34:9036–9047. https://doi.org/10.1016/j.biomaterials.2013.08.011

Murphy JW, Cho Y, Sachpatzidis A, Fan C, Hodsdon ME, Lolis E (2007) Structural and functional basis of CXCL12 (stromal cell-derived factor-1 alpha) binding to heparin. J Biol Chem 282:10018–10027. https://doi.org/10.1074/jbc.M608796200

Newman AC et al (2013) Analysis of stromal cell secretomes reveals a critical role for stromal cell-derived hepatocyte growth factor and fibronectin in angiogenesis. Arterioscler Thromb Vasc Biol 33:513–522. https://doi.org/10.1161/ATVBAHA.112.300782

Nitsche A, Junghahn I, Thulke S, Aumann J, Radonic A, Fichtner I, Siegert W (2003) Interleukin-3 promotes proliferation and differentiation of human hematopoietic stem cells but reduces their repopulation potential in NOD/SCID mice. Stem Cells 21:236–244. https://doi.org/10.1634/stemcells.21-2-236

Oskowitz A, McFerrin H, Gutschow M, Carter ML, Pochampally R (2011) Serum-deprived human multipotent mesenchymal stromal cells (MSCs) are highly angiogenic. Stem Cell Res 6:215–225. https://doi.org/10.1016/j.scr.2011.01.004

Pan S et al (2013) SCF promotes dental pulp progenitor migration, neovascularization, and collagen remodeling - potential applications as a homing factor in dental pulp regeneration. Stem Cell Rev 9:655–667. https://doi.org/10.1007/s12015-013-9442-7

Paquet J, Deschepper M, Moya A, Logeart-Avramoglou D, Boisson-Vidal C, Petite H (2015) Oxygen tension regulates human mesenchymal stem cell paracrine functions. Stem Cells Transl Med 4:809–821. https://doi.org/10.5966/sctm.2014-0180

Pfeffer K (2003) Biological functions of tumor necrosis factor cytokines and their receptors. Cytokine Growth Factor Rev 14:185–191

Pianta S, Signoroni PB, Muradore I, Rodrigues MF, Rossi D, Silini A, Parolini O (2015) Amniotic membrane mesenchymal cells-derived factors skew T cell polarization toward treg and downregulate Th1 and Th17 cells subsets. Stem Cell Rev Rep 11:394–407. https://doi.org/10.1007/s12015-014-9558-4

Pierson W, Liston A (2010) A new role for interleukin-10 in immune regulation. Immunol Cell Biol 88:769–770. https://doi.org/10.1038/icb.2010.105
Pires AO, Neves-Carvalho A, Sousa N, Salgado AJ (2014) The secretome of bone marrow and wharton jelly derived mesenchymal stem cells induces differentiation and neurite outgrowth in SH-SY5Y cells. Stem Cells Int 2014:438352. https://doi.org/10.1155/2014/438352
Pons J et al (2008) VEGF improves survival of mesenchymal stem cells in infarcted hearts. Biochem Biophys Res Commun 376:419–422. https://doi.org/10.1016/j.bbrc.2008.09.003
Ratajczak M, Kucia M, Jadczyk T, Greco N, Wojakowski W, Tendera M, Ratajczak J (2012) Pivotal role of paracrine effects in stem cell therapies in regenerative medicine: can we translate stem cell-secreted paracrine factors and microvesicles into better therapeutic strategies? Leukemia 26:1166–1173
Ribeiro CA, Salgado AJ, Fraga JS, Silva NA, Reis RL, Sousa N (2011) The secretome of bone marrow mesenchymal stem cells-conditioned media varies with time and drives a distinct effect on mature neurons and glial cells (primary cultures). J Tissue Eng Regen Med 5:668–672. https://doi.org/10.1002/term.365
Rojas M, Xu J, Woods CR, Mora AL, Spears W, Roman J, Brigham KL (2005) Bone marrow-derived mesenchymal stem cells in repair of the injured lung. Am J Respir Cell Mol Biol 33:145–152. https://doi.org/10.1165/rcmb.2004-0330OC
Rossi D, Pianta S, Magatti M, Sedlmayr P, Parolini O (2012) Characterization of the conditioned medium from amniotic membrane cells: prostaglandins as key effectors of its immunomodulatory activity. PLoS One 7:e46956. https://doi.org/10.1371/journal.pone.0046956
Rousset F et al (1992) Interleukin 10 is a potent growth and differentiation factor for activated human B lymphocytes. Proc Natl Acad Sci U S A 89:1890–1893. https://doi.org/10.1073/pnas.89.5.1890
Salcedo R et al (1999) Vascular endothelial growth factor and basic fibroblast growth factor induce expression of CXCR4 on human endothelial cells: in vivo neovascularization induced by stromal-derived factor-1alpha. Am J Pathol 154:1125–1135
Salgado A et al (2015) Mesenchymal stem cells secretome as a modulator of the neurogenic niche: basic insights and therapeutic opportunities. Front Cell Neurosci 9. https://doi.org/10.3389/fncel.2015.00249
Sart S, Liu YJ, Ma T, Li Y (2014) Microenvironment regulation of pluripotent stem cell-derived neural progenitor aggregates by human mesenchymal stem cell secretome. Tissue Eng A 20:2666–2679. https://doi.org/10.1089/ten.tea.2013.0437
Schroder K, Hertzog PJ, Ravasi T, Hume DA (2004) Interferon-γ: an overview of signals, mechanisms and functions. J Leukoc Biol 75:163–189. https://doi.org/10.1189/jlb.0603252
Shi Y et al (2006) Granulocyte-macrophage colony-stimulating factor (GM-CSF) and T-cell responses: what we do and don't know. Cell Res 16:126–133. https://doi.org/10.1038/sj.cr.7310017
Son BR et al (2006) Migration of bone marrow and cord blood mesenchymal stem cells in vitro is regulated by stromal-derived factor-1-CXCR4 and hepatocyte growth factor-c-met axes and involves matrix metalloproteinases. Stem Cells 24:1254–1264. https://doi.org/10.1634/stemcells.2005-0271
Stoddart MJ, Bara J, Alini M (2015) Cells and secretome--towards endogenous cell re-activation for cartilage repair. Adv Drug Deliv Rev 84:135–145. https://doi.org/10.1016/j.addr.2014.08.007
Sulpice E et al (2009) Cross-talk between the VEGF-A and HGF signalling pathways in endothelial cells. Biol Cell 101:525–539. https://doi.org/10.1042/BC20080221
Takebayashi T et al (1995) Hepatocyte growth factor/scatter factor modulates cell motility, proliferation, and proteoglycan synthesis of chondrocytes. J Cell Biol 129:1411–1419
Tamama K, Fan VH, Griffith LG, Blair HC, Wells A (2006) Epidermal growth factor as a candidate for ex vivo expansion of bone marrow–derived mesenchymal stem cells. Stem Cells 24:686–695. https://doi.org/10.1634/stemcells.2005-0176
Tamama K, Kawasaki H, Wells A (2010) Epidermal growth factor (EGF) treatment on multipotential stromal cells (MSCs). Possible enhancement of therapeutic potential of MSC. J Biomed Biotechnol 2010:795385. https://doi.org/10.1155/2010/795385

Teixeira FG et al (2015) Secretome of mesenchymal progenitors from the umbilical cord acts as modulator of neural/glial proliferation and differentiation. Stem Cell Rev Rep 11:288–297. https://doi.org/10.1007/s12015-014-9576-2

Teixeira FG et al (2016) Modulation of the mesenchymal stem cell secretome using computer-controlled bioreactors: impact on neuronal cell proliferation, survival and differentiation. Sci Rep 6:27791. https://doi.org/10.1038/srep27791

Teixeira FG et al (2017) Impact of the secretome of human mesenchymal stem cells on brain structure and animal behavior in a rat model of Parkinson's disease. Stem Cells Transl Med 6:634–646. https://doi.org/10.5966/sctm.2016-0071

Wang Y et al (2006) Changes in circulating mesenchymal stem cells, stem cell homing factor, and vascular growth factors in patients with acute ST elevation myocardial infarction treated with primary percutaneous coronary intervention. Heart 92:768–774. https://doi.org/10.1136/hrt.2005.069799

Werner S, Grose R (2003) Regulation of wound healing by growth factors and cytokines. Physiol Rev 83:835–870. https://doi.org/10.1152/physrev.00031.2002

Yanada S, Ochi M, Kojima K, Sharman P, Yasunaga Y, Hiyama E (2006) Possibility of selection of chondrogenic progenitor cells by telomere length in FGF-2-expanded mesenchymal stromal cells. Cell Prolif 39:575–584. https://doi.org/10.1111/j.1365-2184.2006.00397.x

Yew TL et al (2011) Enhancement of wound healing by human multipotent stromal cell conditioned medium: the paracrine factors and p38 MAPK activation. Cell Transplant 20:693–706. https://doi.org/10.3727/096368910X550198

Yu Q, Liu L, Lin J, Wang Y, Xuan X, Guo Y, Hu S (2015) SDF-1α/CXCR4 axis mediates the migration of mesenchymal stem cells to the hypoxic-ischemic brain lesion in a rat model. Cell J (Yakhteh) 16:440–447

Zhang Y et al (2009) A novel function of granulocyte colony-stimulating factor in mobilization of human hematopoietic progenitor cells. Immunol Cell Biol 87:428–432. https://doi.org/10.1038/icb.2009.9

Chapter 3
Preparation of Extracellular Vesicles from Mesenchymal Stem Cells

Fernanda Ferreira Cruz, Ligia Lins de Castro, and Patricia Rieken Macedo Rocco

Introduction

One of the most important mechanisms of paracrine communication between mesenchymal cells (MSCs) occurs by the release of extracellular vesicles (EVs) (Tetta et al. 2013; Ragni et al. 2017). EVs carry proteins, lipids, lnRNAs, mRNAs, and microRNAs, which are capable of reprogramming the phenotypes of other cells (Yuan et al. 2009).

EVs from a variety of sources have shown therapeutic potential, with results often more promising than those obtained with mesenchymal cells themselves (Yuan et al. 2009; Cruz et al. 2015; de Castro et al. 2017). Interest in EVs has been increasing, and different protocols for their collection, processing, and extraction have been published, which has made it difficult to compare different studies on the subject. Therefore, in 2013, the ISEV (International Society of Extracellular Vesicles) published a position paper in an attempt to standardize these protocols (Witwer et al. 2013). These techniques will be discussed below.

Collection and Processing of EVs from MSC Culture Medium

EVs can be collected from the culture medium of MSCs and isolated for characterization and for therapeutic use, but the amount of EVs found in MSCs under normal conditions is generally insufficient for any subsequent application (Rani et al. 2015). One way to optimize the release of EVs is by induction of cellular stress. Several

F. F. Cruz · L. L. de Castro · P. R. M. Rocco (✉)
Laboratory of Pulmonary Investigation, Carlos Chagas Filho Biophysics Institute, Federal University of Rio de Janeiro, Rio de Janeiro, RJ, Brazil
e-mail: prmrocco@biof.ufrj.br

P. V. Pham (ed.), *Stem Cell Drugs - A New Generation of Biopharmaceuticals*, Stem Cells in Clinical Applications, https://doi.org/10.1007/978-3-319-99328-7_3

MSC stress protocols are available, but it must be emphasized that stress causes release of different types and numbers of EVs with different contents than in basal-state cells, and that the different cellular stress protocols used also influence the quantity and type of EVs released (Witwer et al. 2013).

The method most commonly used to induce cell stress is FBS (fetal bovine serum) deprivation (Witwer et al. 2013; de Castro et al. 2017). FBS is an essential requirement for cell culture, as it provides growth factors and vitamins that are needed for cellular growth and expansion (Bieback et al. 2009). Deprived of this supplement, cells cease to proliferate and their viability begins to decrease. Prolonged FSB deprivation induces cytochrome C release, resulting in mitochondrial dysfunction and apoptotic cell death due to lack of nutrients (Zhu et al. 2006; Potier et al. 2007; Wang et al. 2015). However, in another study, MSCs were cultivated in serum-free conditions for a short time and exhibited normal morphology (Fu et al. 2011). These reported discrepancies may be due to differences in the types of cell cultured, the various methods used for FSB deprivation, serum depletion time, and, mainly, differences between the various techniques used to determine cell death or survival (Amiri et al. 2014). According to the ISEV, the maximum acceptable cell death rate for EV extraction without risk of contamination by fragments of dead cells is 5% (Witwer et al. 2013). For EV collection, we maintain MSCs without FBS for 12 h, during which time the cells remain viable and continue to release EVs (unpublished data). This duration of cellular stress due to serum deprivation has been used by several research groups (Monsel et al. 2015; de Castro et al. 2017). Jeppesen et al. (2014) used the Advanced DMEM medium, which does not need to be supplemented with FBS, although supplementation with 1–2% FBS is recommended. In some cell types, viability is increased, while in others, it is decreased. This culture medium can keep human bone marrow-derived MSCs viable, but produced changes in the marker expression pattern in MSCs (Eitan et al. 2015). Serum deprivation may also induce changes in the secretory pattern of MSCs, expressing endothelial-specific proteins (Oskowitz et al. 2011). Chase et al. (2010) cultured mesenchymal cells without any type of serum and supplemented with late-derived growth factor-BB (PDGF-BB), basic fibroblast growth factor (bFGF), and transforming growth factor (TGF)-β, which may be an alternative strategy for subsequent isolation of EVs. Partial FBS depletion is not indicated because this serum contains vesicles with density between 1.09 and 1.16, similar to that of EVs; furthermore, these vesicles contain RNA (Shelke et al. 2014; Eitan et al. 2015). Shelke et al. (2014) centrifuged FBS (pure or with DMEM) for 0, 1.5, or 18 h, added the FBS to the culture medium (10%), and centrifuged the medium at 120,000 × *g*. With 1.5 h of centrifugation, EV depletion was 60% (measured by concentration of RNAs); with 18 h, depletion of EVs was approximately 95%. Besides laborious, this method was not 100% effective.

Many EV isolation protocols involve the removal of FBS and the addition of 0.5% bovine serum albumin (BSA) (Bruno et al. 2012) or 1% human serum albumin (HSA) (Barile et al. 2014) to induce cellular stress. The rationale behind this step is to prevent cell death and thereby reduce the amount of cellular debris and apoptotic bodies that can be released to the conditioned medium. However, it is known that the bovine serum from which albumin is derived contains vesicles, which may have functional effects (Witwer et al. 2013; Shelke et al. 2014), and increased concentrations of EVs have been measured under stressful conditions (Zhang et al. 2012). Other groups reported functional outcomes using EVs isolated from culture media containing 10% fetal calf serum (FCS), where serum was not sufficiently treated to ensure clearance of the EVs (Bian et al. 2014).

Other ways of stimulating the release of EVs include phorbol 12-myristate 13-acetate (PMA) and calcium ionophores, such as ionomycin (Jeppesen et al. 2014). In this case, it is also important to evaluate the viability of the MSCs before initiating the protocol, as there are no reports of the use of these agents in MSC stimulation.

Recently, a group reported that adipose-derived MSCs cultured under exposure to a 0.5-T static magnetic field shed a higher number of extracellular vesicles to the conditioned medium. Additionally, these EVs were richer in growth factors, such as VEGF. Magnetic field exposure might thus be considered an alternative strategy to enhance EV production and effects (Marędziak et al. 2015).

Infected cells can release EVs in different amounts and with distinct composition compared to uninfected cells. In addition, the microorganisms can be unknowingly extracted together with the EVs (Bellingham et al. 2012; Singh et al. 2015). *Mycoplasma*, for example, is 300 nm in size (diameter) and closely resembles EVs when analyzed by scanning electron microscopy (Singh et al. 2015). In addition, the vesicles released by cells contaminated with mycoplasma have an immunosuppressive effect, which may be confused with the effect of healthy MSCs (Quah and O'Neill 2007; Yang et al. 2012). Thus, it is important to confirm that MSCs are mycoplasma-free before starting any experiments.

The amount of EVs collected will depend on the type of MSC chosen, whether EV release by the cell will be stimulated, which stimulus will be used, and the method of extraction, which will be discussed later (Fig. 3.1).

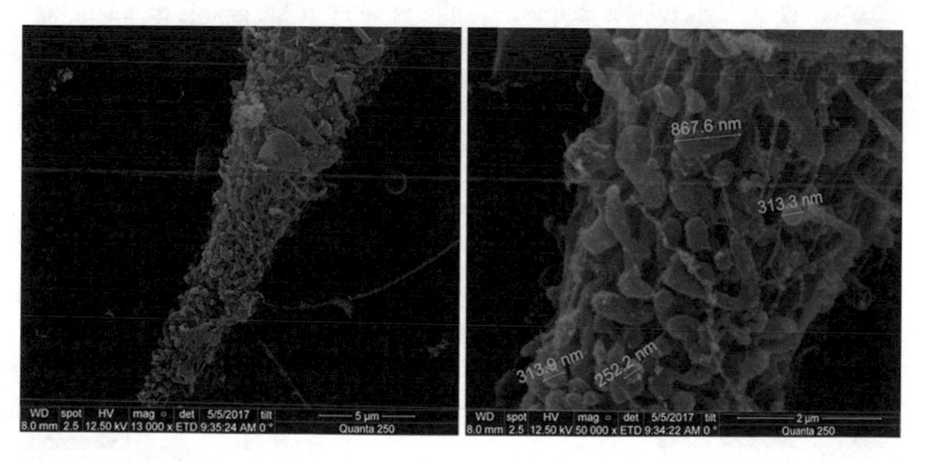

Fig. 3.1 Scanning electron microscopy of MSCs. EVs being released from MSCs after FBS deprivation for 12 h

Methods for EV Collection from MSC Culture Medium

Differential Centrifugation

Differential centrifugation is the most widely used method and is the gold standard for EVs (Sáenz-Cuesta et al. 2015). It consists of the separation of particles by a centrifugation sequence, taking into account that these particles have different sedimentation coefficients (Rickwood et al. 1994). To determine the coefficient of sedimentation of a particle, the following formula is applied:

$$S = \frac{m}{6\pi\eta r},$$

where S denotes the settling coefficient, m is the mass of the particle, η denotes the viscosity of the medium, and r is the shape of the particle. The sedimentation coefficient tells us how fast a particle sediments; larger particles sediment first and thus have a higher coefficient of sedimentation than smaller particles (Rickwood et al. 1994). High viscosity leads to lower sedimentation efficiency (Momen-Heravi et al. 2012). Table 3.1 lists some viscosity values that can be used in the formula above (Momen-Heravi et al. 2012).

First, a rapid centrifugation at 2000 × *g* is done to remove cellular debris and possible apoptotic bodies, and the resulting pellet is discarded. The supernatant is centrifuged more slowly so that the smaller vesicles can be isolated. For isolation of microvesicles, a rotation of 10,000–20,000 × *g* is required (Ismail et al. 2013; Witwer et al. 2013; Cvjetkovic et al. 2014). A study of EVs from umbilical cord-derived MSCs reported that centrifugation at 40,000 × *g* is already capable of contaminating the sample of microvesicles with exosomes (Rad et al. 2016). For isolation of exosomes, the supernatant from this first centrifugation should be centrifuged again at 100,000 × *g* or higher, since they are smaller vesicles. If the goal is to obtain both populations of EVs, centrifugation can be performed at rates from 2000 × *g* to 1,000,000 × *g* (Ismail et al. 2013; Witwer et al. 2013; Cvjetkovic et al. 2014).

Another important point is the duration of centrifugation. If the first centrifugation is done very quickly, larger particles such as cell debris and apoptotic bodies will remain in the supernatant and will be decanted in the next centrifugation along with the EVs. Overly slow centrifugation can sediment the EVs of interest, causing them to be discarded. The following ultracentrifugation runs should also be performed at the optimum time so that the amount of isolated EVs is sufficient (Cvjetkovic et al. 2014).

Centrifugation speed and centrifugation time are essential for the desired result. Two types of rotors are used in ultracentrifuges: the fixed-angle rotor and the swinging bucket rotor (Cvjetkovic et al. 2014; Livshts et al. 2015).

Table 3.1 Viscosity values of the main fluids used to extract EVs

Fluid	Viscosity
FBS	1.4
Culture medium	1.1
PBS	1

With a fixed-angle rotor, the tubes are kept at a fixed angle in the rotor cavity, and at the end of centrifugation, the sediment containing the EVs remains on the side of the tube facing the outside of the centrifuge. With a swinging bucket rotor, the tubes are allocated to the rotor that is at rest, while the samples swing vertically; the EVs thus collect at the bottom of the tube at the end of the centrifugation process (Rickwood et al. 1994; Livshts et al. 2015).

The choice of rotor depends on several factors. First, one must consider the g-force required for the type of EV to be extracted and the volume to be placed into the tubes. Fixed rotors are used most often when differential centrifugation is required because the EVs form the pellet by a shorter path, and decantation is thus faster and more efficient. Swinging rotors allow better individual separation of EVs, favoring centrifugal gradient separation, since pellet formation is relatively inefficient (Livshts et al. 2015).

The larger the sample volume, the larger the rotor required, and the larger the rotor, the lower its rotation speed. To calculate the required velocity, one must know the clearing factor (k), which is the efficiency of pellet formation at maximum velocity. The closer the clearing factor is to zero, the greater the efficiency of pellet formation. It is also important to calculate the speed difference between one rotor and another when needing to switch to a different rotor while keeping the same pellet characteristics (Rickwood et al. 1994; Livshts et al. 2015). The clearing factor is expressed as (Rickwood et al. 1994; Livshts et al. 2015):

$$k = \frac{2.53 \times 10^5 \times \text{In}\left(\frac{\text{rmax}}{\text{rmin}}\right)}{\left(\frac{\text{rpm}}{1000}\right)^2},$$

the maximal radius (rmax) and minimal (rmin) radius are supplied by the rotor manufacturer, as is the maximum rpm. Some of the most commonly used rotors and their clearing factors are described in Table 3.2.

The time of pellet formation can also be calculated (Rickwood et al. 1994; Cvjetkovic et al. 2014):

$$T = \frac{k}{S},$$

where T denotes the time in hours, k is the clearing factor, and S denotes the coefficient of sedimentation, as mentioned above.

The difference in time from one rotor to another can be calculated as long as the content to be centrifuged is the same (Rickwood et al. 1994; Cvjetkovic et al. 2014):

Table 3.2 Rotors most commonly used for EV extraction and their clearing factors (Beckman Coulter)

70 Ti	$k = 44$
45 Ti	$k = 133$
SW 32 Ti	$k = 204$

Table 3.3 Time equivalences between the fixed and swinging rotors most used for EV extraction 118,000 × *g*

Rotor	RPM	Equivalent time (min)
70 Ti	40,045	70
45 Ti	38,837	93
SW41 Ti	30,913	114
SW32 Ti	30,998	114
TLA-100.3	52,724	27

$$\frac{T1}{k1} = \frac{T2}{k2},$$

where T denotes the time in hours, k is the clearing factor, and 1 and 2 denote the different rotors.

These calculations can be performed on manufacturers' websites, as long as both rotors are from the same company. Beckman Coulter, for example, allows rotor 1 to be selected within a list of rotors. By setting the rpm and centrifugation time, the equivalent time for the second rotor can be calculated.

The above calculations only work for rotors of the same type. Cvjetkovic et al. (2014) calculated the equivalence between the fixed and swinging rotors most often used for EV extraction at 118,000 × g (Table 3.3).

Differential centrifugation is limited by protein contamination. Washing the pellet with PBS and performing a new ultracentrifuge run after this wash may reduce protein contamination, but can eliminate desired components (Franquesa et al. 2014; Conforti et al. 2014). These differential ultracentrifugation procedures are not efficient for size separation, because sedimentation also depends on the density or "charge" of a particle and the distance it travels. Some small EVs near the bottom of the tube will sediment along with large particles even at low speed, while some larger particles at the top of the tube can sediment only with high-speed rotation. Aggregation of EVs is a common occurrence, and also affects the separation of individual vesicles (Ismail et al. 2013; Witwer et al. 2013).

Different protocols for EV extraction by differential centrifugation lead to inconsistencies in the isolated material, which may explain the different biological effects of EV from MSCs reported by different research groups (Bian et al. 2014; Conforti et al. 2014).

Immunoaffinity Isolation

Immunoaffinity isolation is based on the presence of specific surface markers in EV subpopulations (Clayton et al. 2001; Wubbolts et al. 2003; Théry et al. 2006). Antibodies to surface proteins are used to positively select the desired EV populations (immunoselection) or to capture unwanted EV populations (negative selection or immunodepletion) (Yoo et al. 2008; Mathivanan et al. 2010; Kim et al. 2012). Antibodies are combined with beads or other matrices, and, by covalent or high-affinity interactions, facilitate physical separation by low-speed centrifugation or magnetic techniques. Depending on the approach, this method can be used to purify and enrich EVs (Witwer et al. 2013).

Because this technique has high specificity (Rana et al. 2012; Tauro et al. 2012), it is used when only one subpopulation is desired. It is important to be aware that some markers used in the selection of EVs may not be present or recognized in all EVs, which leads to a yield much lower than with methods that extract EVs based on physical characteristics. Optimally, EVs should be evaluated not only for the presence of selected markers, but also for the absence of markers that are not of interest, including appropriate isotype controls (Witwer et al. 2013).

Density Gradient

The density gradient method is based on size and density. It is usually combined with ultracentrifugation. Two types of devices are available, which differ in sample loading position: top-loading and bottom-loading (Choi et al. 2011; Willms et al. 2016).

In top-loading devices, the high-density particles are at the bottom and the low-density particles at the top. Samples are placed at the top of the tubes, and visible particle separation occurs after centrifugation. In this method, separation depends more on the size and mass of the particles than on their density. If particles of different sizes and the same density are centrifuged for a long enough time, they may eventually be in the same position. Since prolonged centrifugation can sediment the smallest particles, it is important to determine the optimal centrifugation time (Choi et al. 2011; Willms et al. 2016).

The bottom-loading method is based on particle density. Higher-density particles remain at the bottom of the tubes after centrifugation, at which time the particles are in an equal density gradient medium. Size affects only the velocity of particle motion until the density of the particle is equal to that of the density gradient of the medium, also called the velocity of flotation (Choi et al. 2011; Willms et al. 2016).

The gradient media used are composed of sucrose and iodixanol (Choi et al. 2011; Willms et al. 2016). The density of the sucrose solution will depend on its osmolarity. The density of iodixanol-based medium, also known as OptiPrep gradient, varies according to the concentration of iodixanol in the purchased solution. The OptiPrep datasheet notes that the solution is isosmotic and has low viscosity, which does not affect EVs, unlike sucrose solution, which is hyperosmotic and high-viscosity and may thus affect EV functionality (Progen 2017).

Van Deun et al. (2014) compared several methods of extracting EVs and concluded that the density gradient method yields the least EVs, but is optimal when the purity of the EVs is more important than their quantity.

To use the density gradient method without separating EV subpopulations, one can perform the cushion-based isolation method, which consists of the use of two gradients: a high-density background gradient composed of 2.5 M sucrose and 50% OptiPrep and a low-density gradient composed of 0.5 M sucrose and 10% OptiPrep. The sample is placed at the top of the tube. After centrifugation, the large particles will remain between the two gradients and the EVs will remain at the top of the tube. One limitation of this method is that contaminant proteins can remain together with the EVs at the end of centrifugation; however, samples may be previously concentrated so as to be free of these proteins (Lamparski et al. 2002; Choi et al. 2007)

Using the density gradient method, Haga et al. (2017) extracted EVs from human and murine bone marrow-derived MSCs with a size of 116 ± 46 nm and 112 ± 56 nm, respectively. Therapeutic use of these EVs was effective in a murine model of lethal hepatic injury. Collino et al. (2017) also isolated EVs derived from human bone marrow MSCs by the density gradient method and obtained size peaks between 100 and 180 nm. These EVs were successfully used in an ischemia-reperfusion renal injury model.

Size Exclusion Filter

In the size exclusion filter method, particles larger than the desired size may be excluded, for example, with a pore size filter of 0.8 mm, or particles smaller than the desired size range can be removed while the target population is maintained in the filter. This method does not enrich EV subpopulations, unless low-molecular-weight filters are used to concentrate the desired populations. However, EVs may stick to the filter and be lost. An alternative is to combine this method with ultracentrifugation or other techniques. Researchers often use 0.8-mm filters to remove large cell fragments prior to EV isolation, while 0.2-mm filters can be used when smaller EVs are desired (Théry et al. 2006; György et al. 2011; Witwer et al. 2013; Franquesa et al. 2014)

Forcing the particles through the pores of the filter can cause deformation and dissolution of large vesicles, compromising their utility. An alternative would be size exclusion by gravity, which can be very time-consuming and may be impracticable. It is therefore advisable to apply as little force as possible to the filter and check that the filters do not release contaminating particles that may interfere with the final results of experiments (Livshts et al. 2015).

Franquesa et al. (2014) isolated EVs from MSCs derived from human adipose tissue by this method. After low-speed centrifugation, 0.2-μm pore filters were run through the sample under pressure, and then combined with ultracentrifugation to isolate EVs with a peak size of 115 nm.

Size Exclusion Chromatography

This method separates the particles dissolved in the medium, based on their size, by pumping the fluid through columns containing gel micropores. Detectors assess light scattering, concentration, and viscosity in the medium. Large particles do not enter the gel and are excluded, while smaller particles enter and can be analyzed (Böing et al. 2014). This method is very effective in obtaining purified EVs, i.e., without contaminating proteins, which favors its use in proteomic analysis. Another advantage is that EVs do not form aggregates. It is a rapid method, but suboptimal if the goal is to obtain EVs in large quantities (Momen-Heravi et al. 2012; Böing et al. 2014; Nordin et al. 2015).

Kim et al. (2015) used this method on human MSCs and obtained EVs with size between 209 ± 1.8 nm and 231 ± 3.2 nm, which were successfully used in an animal model of traumatic brain injury, with a beneficial effect on cognitive recovery.

Kit-Based Precipitation

Some kits for EV isolation are commercially available. The kits are based on volume-excluding polymers, specifically polyethylene glycol (PEG). The most widely used kits are Total Exosome Isolation (Life Technologies), Exoquick (System Biosciences), and Exoprep (Hansabiomed). Methods of isolating EVs using organic solvents such as acetate buffer and acetone (protein organic solvent precipitation, PROSPR) have also been used (Van Deun et al. 2014; Gallart-Palau et al. 2015).

PEG is nontoxic and soluble in water and is the most efficient polymer for EV precipitation. In brief, the culture medium is incubated overnight in the precipitation solution and a low-spin centrifugation is performed to precipitate the EVs. The PEG takes the place of the culture medium, which concentrates until it exceeds its solubility and precipitates (Van Deun et al. 2014).

PEG-extracted EVs from MSCs have been used effectively in treatment-refractory graft-versus-host disease (Kordelas et al. 2014)

Van Deun et al. (2014) compared two of the three kits mentioned above versus ultracentrifugation (differential centrifugation) and OptiPrep and reached the following conclusions, which are extremely useful for those who are seeking to start EV extraction but do not know how to choose (Table 3.4).

The acetate buffer neutralizes the EVs, which are negatively charged because of the presence of phosphatidylserine, promoting hydrophobic interactions and resulting in aggregation and precipitation of the EVs. This method has similar results to ultracentrifugation in terms of the quantity and morphology of the EVs; however, some researchers have observed that soluble proteins end up precipitating along with the EVs (Brownlee et al. 2014).

Table 3.4 Comparisons among EV isolation methods (ultracentrifugation, OptiPrep density gradient, Exoquick kit, and Total Exosome Isolation kit), according to purity, exosome yield, protein yield, RNA yield, ease-of-use, turnaround time, hands-on time, and cost

	Ultracentrifugation	OptiPrep	Exoquick kit	Total exosome isolation kit
Purity	Moderate	Very high	Low	Low
Exosome yield	High	Very high	Low	Low
Protein yield	Moderate	Low	High	Very high
RNA yield	High	Low	High	Moderate
Ease-of-use	Moderate	Low	High	High
Turnaround time (h)	4	20	13	13
Hands-on time (h)	<1	1	<0.5	<0.5
Cost (€)	5	15	15	5

Modified from Van Deun et al. 2014

The above methods may precipitate protein aggregates and are ideally evaluated by other methods, such as the density gradient technique. Also, as in all methods mentioned above, the isolated content should be characterized to make sure that the desired population is being used (Van Deun et al. 2014).

PROSPR used acetone, which removes the soluble proteins from the medium, thus facilitating future analyses (Gallart-Palau et al. 2015).

EV kits are inexpensive, easy to use, suitable for small and large volumes of EVs, and are an excellent alternative for those who do not have equipment such as an ultracentrifuge and chromatography system.

Storage of EVs

A major challenge when working with EVs is the marked impact that preanalytical treatment has on the outcome of analysis. Many investigators have highlighted the importance of a consistent protocol for sample collection and EV preparation, as mentioned before (Lacroix et al. 2012). Factors related to storage, such as freezing temperature and time, freeze–thaw cycles, and transportation, have been examined as well (Bæk et al. 2016).

It may be advisable to proceed to vesicle isolation immediately after collecting the biofluid or cell-conditioned medium that is to be analyzed. In some cases, however, fluid storage before EV purification may be convenient, for instance to allow simultaneous processing of samples from different patients or sources, or when wishing to examine previously biobanked patient samples. At a minimum, cells and platelets should be removed from the fluid prior to storage (Witwer et al. 2013).

The impact of storage has not been methodically evaluated in MSC-conditioned media, but has been evaluated in plasma samples from healthy donors. First, the effect of a single cycle of freezing at −80 °C followed by thawing 1 week later was evaluated. Only a limited increase in EV counts and procoagulant activity was observed. Second, the impact of storage delay was evaluated between 1 week and

1 year. In these conditions, no major change in EV counts or size was observed after 12 months at −80 °C. Third, flash-freezing in liquid nitrogen before storage at −80 °C was compared with a direct −80 °C freezing procedure, and no significant change was observed. Finally, the impact of thawing conditions was evaluated, comparing thawing at 37 °C in a water bath, at room temperature, and on ice. Thawing strategies had no impact on EV size or number, but thawing at room temperature resulted in a significant increase in thrombin generation. Altogether, these results suggest that freeze-thawing and storage conditions can strongly influence EV analysis when performed adequately (Lacroix et al. 2012).

In parallel, storage of granulocyte-derived EVs has been studied. Storage at +20 °C or +4 °C resulted in a significant decrease in EV counts and antibacterial effect after 1 day. Storage at −20 °C did not influence EV counts up to 28 days, but did induce a shift in EV size and almost complete loss of antibacterial function by 28 days. Storage at −80 °C had no significant effect on EV number or size, and allowed partial preservation of the antibacterial function up to 28 days. Flash freezing did not improve these results, whereas the widely used cryoprotectants dimethyl sulfoxide (DMSO, 1%) and glycerin (5%) fully or partially lysed the EVs. Storage significantly altered both the physical and functional properties of EVs, even when the number of EVs remained constant. Thus, if storage is needed, EVs should be kept at −80 °C and preferably for no longer than 7 days. For functional tests, freshly prepared EVs are recommended (Lőrincz et al. 2014). On the other hand, biologic activity was seen in freshly isolated vesicles and in vesicles derived from MSCs stored for up to 6 months in 10% DMSO at −80 °C when used to reverse radiation damage to bone-marrow stem cells (Wen et al. 2016). Recently, new cryopreservation methods have been tested as well. For instance, trehalose, a natural, nontoxic sugar widely used as a protein stabilizer and cryoprotectant by the food and drug industry, was tested as a pancreatic beta-cell exosome-like vesicle storage buffer and found to narrow the particle size distribution; there were no signs of lysis or incomplete vesicles on cryo-electron tomography, and biological activity was preserved (Bosch et al. 2016).

There is still little consensus regarding storage of MSC-derived EVs. Further well-controlled experiments are needed to elucidate the impact of storage temperature and duration, use of cryopreservants, and thawing methods on EV phenotype and recovery (Witwer et al. 2013).

References

Amiri F, Halabian R, Salimian M et al (2014) Induction of multipotency in umbilical cord-derived mesenchymal stem cells cultivated under suspension conditions. Cell Stress Chaperones 19:657–666. https://doi.org/10.1007/s12192-014-0491-x

Bæk R, Søndergaard EKL, Varming K, Jørgensen MM (2016) The impact of various preanalytical treatments on the phenotype of small extracellular vesicles in blood analyzed by protein microarray. J Immunol Methods 438:11–20. https://doi.org/10.1016/j.jim.2016.08.007

Barile L, Lionetti V, Cervio E et al (2014) Extracellular vesicles from human cardiac progenitor cells inhibit cardiomyocyte apoptosis and improve cardiac function after myocardial infarction. Cardiovasc Res 103:530–541. https://doi.org/10.1093/cvr/cvu167

Bellingham SA, Coleman BM, Hill AF (2012) Small RNA deep sequencing reveals a distinct miRNA signature released in exosomes from prion-infected neuronal cells. Nucleic Acids Res 40:10937–10949. https://doi.org/10.1093/nar/gks832

Bian S, Zhang L, Duan L et al (2014) Extracellular vesicles derived from human bone marrow mesenchymal stem cells promote angiogenesis in a rat myocardial infarction model. J Mol Med (Berl) 92:387–397. https://doi.org/10.1007/s00109-013-1110-5

Bieback K, Hecker A, Kocaömer A et al (2009) Human alternatives to fetal bovine serum for the expansion of mesenchymal stromal cells from bone marrow. Stem Cells 27:2331–2341. https://doi.org/10.1002/stem.139

Böing AN, Van Der Pol E, Grootemaat AE et al (2014) Single-step isolation of extracellular vesicles from plasma by size-exclusion chromatography. J Extracell Vesicles 3:118. https://doi.org/10.3402/jev.v3.23430

Bosch S, de Beaurepaire L, Allard M et al (2016) Trehalose prevents aggregation of exosomes and cryodamage. Sci Rep 6:36162. https://doi.org/10.1038/srep36162

Brownlee Z, Lynn KD, Thorpe PE, Schroit AJ (2014) A novel "salting-out" procedure for the isolation of tumor-derived exosomes. J Immunol Methods 407:120–126. https://doi.org/10.1016/j.jim.2014.04.003

Bruno S, Grange C, Collino F et al (2012) Microvesicles derived from mesenchymal stem cells enhance survival in a lethal model of acute kidney injury. PLoS One 7:e33115. https://doi.org/10.1371/journal.pone.0033115

Chase LG, Lakshmipathy U, Solchaga LA et al (2010) A novel serum-free medium for the expansion of human mesenchymal stem cells. Stem Cell Res Ther 1:8. https://doi.org/10.1186/scrt8

Choi D-S, Lee J-M, Park GW et al (2007) Proteomic analysis of microvesicles derived from human colorectal cancer cells. J Proteome Res 6:4646–4655. https://doi.org/10.1021/pr070192y

Choi D-S, Park JO, Jang SC et al (2011) Proteomic analysis of microvesicles derived from human colorectal cancer ascites. Proteomics 11:2745–2751. https://doi.org/10.1002/pmic.201100022

Clayton A, Court J, Navabi H et al (2001) Analysis of antigen presenting cell derived exosomes, based on immuno-magnetic isolation and flow cytometry. J Immunol Methods 247:163–174. https://doi.org/10.1016/S0022-1759(00)00321-5

Collino F, Pomatto M, Bruno S et al (2017) Exosome and microvesicle-enriched fractions isolated from mesenchymal stem cells by gradient separation showed different molecular signatures and functions on renal tubular epithelial cells. Stem Cell Rev 13:226–243. https://doi.org/10.1007/s12015-016-9713-1

Conforti A, Scarsella M, Starc N et al (2014) Microvescicles derived from mesenchymal stromal cells are not as effective as their cellular counterpart in the ability to modulate immune responses in vitro. Stem Cells Dev 23:2591–2599. https://doi.org/10.1089/scd.2014.0091

Cruz FF, Borg ZD, Goodwin M et al (2015) Systemic administration of human bone marrow-derived mesenchymal stromal cell extracellular vesicles ameliorates aspergillus hyphal extract-induced allergic airway inflammation in immunocompetent mice. Stem Cells Transl Med 4:1302. https://doi.org/10.5966/sctm.2014-0280

Cvjetkovic A, Lötvall J, Lässer C (2014) The influence of rotor type and centrifugation time on the yield and purity of extracellular vesicles. J Extracell Vesicles 3:3–4. https://doi.org/10.3402/jev.v3.23111

de Castro LL, Xisto DG, Kitoko JZ et al (2017) Human adipose tissue mesenchymal stromal cells and their extracellular vesicles act differentially on lung mechanics and inflammation in experimental allergic asthma. Stem Cell Res Ther 8:151. https://doi.org/10.1186/s13287-017-0600-8

Eitan E, Zhang S, Witwer KW, Mattson MP (2015) Extracellular vesicle-depleted fetal bovine and human sera have reduced capacity to support cell growth. J Extracell Vesicles 4:1–10. https://doi.org/10.3402/jev.v4.26373

Franquesa M, Hoogduijn MJ, Ripoll E et al (2014) Update on controls for isolation and quantification methodology of extracellular vesicles derived from adipose tissue mesenchymal stem cells. Front Immunol 5. https://doi.org/10.3389/fimmu.2014.00525

Fu WL, Jia ZQ, Wang WP et al (2011) Proliferation and apoptosis property of mesenchymal stem cells derived from peripheral blood under the culture conditions of hypoxia and serum deprivation. Chin Med J 124:3959–3967. https://doi.org/10.3760/cma.j.issn.0366-6999.2011.23.022

Gallart-Palau X, Serra A, Wong ASW et al (2015) Extracellular vesicles are rapidly purified from human plasma by PRotein organic solvent PRecipitation (PROSPR). Sci Rep 5:14664. https://doi.org/10.1038/srep14664

György B, Módos K, Pállinger É et al (2011) Detection and isolation of cell-derived microparticles are compromised by protein complexes resulting from shared biophysical parameters. Blood 117:e39. https://doi.org/10.1182/blood-2010-09-307595

Haga H, Yan IK, Takahashi K et al (2017) Extracellular vesicles from bone marrow-derived mesenchymal stem cells improve survival from lethal hepatic failure in mice. Stem Cells Transl Med 6:1262–1272. https://doi.org/10.1002/sctm.16-0226

Ismail N, Wang Y, Dakhlallah D et al (2013) Macrophage microvesicles induce macrophage differentiation and miR-223 transfer. Blood 121:984–995. https://doi.org/10.1182/blood-2011-08-374793

Jeppesen DK, Hvam ML, Primdahl-Bengtson B et al (2014) Comparative analysis of discrete exosome fractions obtained by differential centrifugation. J Extracell Vesicles 3. https://doi.org/10.3402/jev.v3.25011

Kim G, Yoo CE, Kim M et al (2012) Noble polymeric surface conjugated with zwitterionic moieties and antibodies for the isolation of exosomes from human serum. Bioconjug Chem 23:2114–2120. https://doi.org/10.1021/bc300339b

Kim D-K, Nishida H, An SY et al (2015) Chromatographically isolated CD63+CD81+ extracellular vesicles from mesenchymal stromal cells rescue cognitive impairments after TBI. Proc Natl Acad Sci U S A 113:170. https://doi.org/10.1073/pnas.1522297113

Kordelas L, Rebmann V, Ludwig A et al (2014) MSC-derived exosomes: a novel tool to treat therapy-refractory graft-versus-host disease. Leukemia 28:970–973. https://doi.org/10.1038/leu.2014.41

Lacroix R, Judicone C, Poncelet P et al (2012) Impact of pre-analytical parameters on the measurement of circulating microparticles: towards standardization of protocol. J Thromb Haemost 10:437–446. https://doi.org/10.1111/j.1538-7836.2011.04610.x

Lamparski HG, Metha-Damani A, Yao JY et al (2002) Production and characterization of clinical grade exosomes derived from dendritic cells. J Immunol Methods 270:211–226. https://doi.org/10.1016/S0022-1759(02)00330-7

Livshts MA, Khomyakova E, Evtushenko EG et al (2015) Isolation of exosomes by differential centrifugation: theoretical analysis of a commonly used protocol. Sci Rep 5:17319. https://doi.org/10.1038/srep17319

Lőrincz ÁM, Timár CI, Marosvári KA et al (2014) Effect of storage on physical and functional properties of extracellular vesicles derived from neutrophilic granulocytes. J Extracell Vesicles 3:25465. https://doi.org/10.3402/jev.v3.25465\r25465

Marędziak M, Marycz K, Lewandowski D et al (2015) Static magnetic field enhances synthesis and secretion of membrane-derived microvesicles (MVs) rich in VEGF and BMP-2 in equine adipose-derived stromal cells (EqASCs)-a new approach in veterinary regenerative medicine. In Vitro Cell Dev Biol Anim 51:230–240. https://doi.org/10.1007/s11626-014-9828-0

Mathivanan S, Lim JWE, Tauro BJ et al (2010) Proteomics analysis of A33 immunoaffinity-purified exosomes released from the human colon tumor cell line LIM1215 reveals a tissue-specific protein signature. Mol Cell Proteomics 9:197–208. https://doi.org/10.1074/mcp.M900152-MCP200

Momen-Heravi F, Balaj L, Alian S et al (2012) Impact of biofluid viscosity on size and sedimentation efficiency of the isolated microvesicles. Front Physiol 3. https://doi.org/10.3389/fphys.2012.00162

Monsel A, Zhu YG, Gennai S et al (2015) Therapeutic effects of human mesenchymal stem cell-derived microvesicles in severe pneumonia in mice. Am J Respir Crit Care Med 192:324. https://doi.org/10.1164/rccm.201410-1765OC

Nordin JZ, Lee Y, Vader P et al (2015) Ultrafiltration with size-exclusion liquid chromatography for high yield isolation of extracellular vesicles preserving intact biophysical and functional properties. Nanomed Nanotechnol Biol Med 11:879–883. https://doi.org/10.1016/j.nano.2015.01.003
Oskowitz A, McFerrin H, Gutschow M et al (2011) Serum-deprived human multipotent mesenchymal stromal cells (MSCs) are highly angiogenic. Stem Cell Res 6:215–225. https://doi.org/10.1016/j.scr.2011.01.004
Potier E, Ferreira E, Meunier A et al (2007) Prolonged hypoxia concomitant with serum deprivation induces massive human mesenchymal stem cell death. Tissue Eng 13:1325–1331. https://doi.org/10.1089/ten.2006.0325
Progen OptiPrep datasheet (2017). https://www.progen.com/media/downloads/datasheets/1114542.pdf. Accessed 8 Jan 2017
Quah BJ, O'Neill HC (2007) Mycoplasma contaminants present in exosome preparations induce polyclonal B cell responses. J Leukoc Biol 82:1070–1082. jlb.0507277 [pii]. https://doi.org/10.1189/jlb.0507277
Rad F, Pourfathollah AA, Yari F et al (2016) Microvesicles preparation from mesenchymal stem cells. Med J Islam Repub Iran 30:398
Ragni E, Banfi F, Barilani M et al (2017) Extracellular vesicle-shuttled mRNA in mesenchymal stem cell communication. Stem Cells 35:1093–1105. https://doi.org/10.1002/stem.2557
Rana S, Yue S, Stadel D, Zöller M (2012) Toward tailored exosomes: the exosomal tetraspanin web contributes to target cell selection. Int J Biochem Cell Biol 44:1574–1584. https://doi.org/10.1016/j.biocel.2012.06.018
Rani S, Ryan AE, Griffin MD, Ritter T (2015) Mesenchymal stem cell-derived extracellular vesicles: toward cell-free therapeutic applications. Mol Ther 23:812–823. https://doi.org/10.1038/mt.2015.44
Rickwood D, Ford T, Steensgaard J (1994) Centrifugation: essential data. Wiley, Chichester, UK
Sáenz-Cuesta M, Arbelaiz A, Oregi A et al (2015) Methods for extracellular vesicles isolation in a hospital setting. Front Immunol 6. https://doi.org/10.3389/fimmu.2015.00050
Shelke GV, Lässer C, Gho YS, Lötvall J (2014) Importance of exosome depletion protocols to eliminate functional and RNA-containing extracellular vesicles from fetal bovine serum. J Extracell Vesicles 3. https://doi.org/10.3402/jev.v3.24783
Singh PP, Li L, Schorey JS (2015) Exosomal RNA from mycobacterium tuberculosis-infected cells is functional in recipient macrophages. Traffic 16:555–571. https://doi.org/10.1111/tra.12278
Tauro BJ, Greening DW, Mathias RA et al (2012) Comparison of ultracentrifugation, density gradient separation, and immunoaffinity capture methods for isolating human colon cancer cell line LIM1863-derived exosomes. Methods 56:293–304. https://doi.org/10.1016/j.ymeth.2012.01.002
Tetta C, Ghigo E, Silengo L et al (2013) Extracellular vesicles as an emerging mechanism of cell-to-cell communication. Endocrine 44:11–19
Théry C, Amigorena S, Raposo G, Clayton A (2006) Isolation and characterization of exosomes from cell culture supernatants and biological fluids. Curr Protoc Cell Biol Chapter 3:Unit 3.22. https://doi.org/10.1002/0471143030.cb0322s30, 30, 3.22.1
Van Deun J, Mestdagh P, Sormunen R et al (2014) The impact of disparate isolation methods for extracellular vesicles on downstream RNA profiling. J Extracell Vesicles 3. https://doi.org/10.3402/jev.v3.24858
Wang F, Zhou H, Du Z et al (2015) Cytoprotective effect of melatonin against hypoxia/serum deprivation-induced cell death of bone marrow mesenchymal stem cells in vitro. Eur J Pharmacol 748:157–165. https://doi.org/10.1016/j.ejphar.2014.09.033
Wen S, Dooner M, Cheng Y et al (2016) Mesenchymal stromal cell-derived extracellular vesicles rescue radiation damage to murine marrow hematopoietic cells. Leukemia 30:2221–2231. https://doi.org/10.1038/leu.2016.107
Willms E, Johansson HJ, Mäger I et al (2016) Cells release subpopulations of exosomes with distinct molecular and biological properties. Sci Rep 6:22519. https://doi.org/10.1038/srep22519

Witwer KW, Buzás EI, Bemis LT et al (2013) Standardization of sample collection, isolation and analysis methods in extracellular vesicle research. J Extracell Vesicles 2:1–25. https://doi.org/10.3402/jev.v2i0.20360

Wubbolts R, Leckie RS, Veenhuizen PTM et al (2003) Proteomic and biochemical analyses of human B cell-derived exosomes: potential implications for their function and multivesicular body formation. J Biol Chem 278:10963–10972. https://doi.org/10.1074/jbc.M207550200

Yang C, Chalasani G, Ng YH, Robbins PD (2012) Exosomes released from mycoplasma infected tumor cells activate inhibitory B cells. PLoS One 7:e36138. https://doi.org/10.1371/journal.pone.0036138

Yoo PB, Woock JP, Grill WM (2008) Somatic innervation of the feline lower urinary tract. Brain Res 1246:80–87. https://doi.org/10.1016/j.brainres.2008.09.053

Yuan A, Farber EL, Rapoport AL et al (2009) Transfer of microRNAs by embryonic stem cell microvesicles. PLoS One 4:e4722. https://doi.org/10.1371/journal.pone.0004722

Zhang H-C, Liu X-B, Huang S et al (2012) Microvesicles derived from human umbilical cord mesenchymal stem cells stimulated by hypoxia promote angiogenesis both in vitro and in vivo. Stem Cells Dev 21:3289–3297. https://doi.org/10.1089/scd.2012.0095

Zhu W, Chen J, Cong X et al (2006) Hypoxia and serum deprivation-induced apoptosis in mesenchymal stem cells. Stem Cells 24:416–425. https://doi.org/10.1634/stemcells.2005-0121

Chapter 4
Exosomes for Regeneration, Rejuvenation, and Repair

Joydeep Basu and John W. Ludlow

Introduction

The evolution in biological pharmaceuticals has been defined by a paradigm shift away from application leveraging of the cell as a simple manufacturing platform for therapeutically relevant proteins and toward an appreciation of the cell itself as the active biological ingredient for mediating regeneration and repair of diseased tissue. More recently, methodologies for application of stem and progenitor cell populations for tissue engineering and regenerative medicine are being significantly impacted by the growing recognition that the action of secreted cell-derived byproducts functioning at a distance, as opposed to site-specific integration and directed differentiation, is the salient mechanism of action by which these cell populations catalyze regenerative outcomes (Basu and Ludlow 2014; Guthrie et al. 2013). This *secretome* is composed of a regenerative milieu largely consisting of proteins, nucleic acids and membrane-bound vesicles of a range of sizes that are potentially able to independently triggering regeneration and repair as well as catalyzing the de novo organogenesis of tissue engineered organs ex vivo (Maguire 2013; Justewicz et al. 2012). These results highlight a transitional return toward leveraging the cell as a medicinal factory with the secretome rather than the cell itself now representing the active biological ingredient (Caplan and Correa 2011). In this chapter, we review the relevant recent literature and examine this paradigm shift away from the manufacture and application of cells toward cell-derived regenerative by-products such as exosomes. Specific examples documenting regeneration of heart, kidney, skin,

J. Basu (✉)
Twin City Bio, LLC, Winston-Salem, NC, USA
e-mail: jbasu@twincitybio.com

J. W. Ludlow
Zen-Bio, Inc., Research Triangle Park, NC, USA

P. V. Pham (ed.), *Stem Cell Drugs - A New Generation of Biopharmaceuticals*, Stem Cells in Clinical Applications, https://doi.org/10.1007/978-3-319-99328-7_4

and tendon through application of exosome-based therapies will be evaluated. We will leverage lessons learned from cell-based systems to illustrate process development, scale-up, manufacturing, quality control, regulatory, and intellectual property issues associated with exosome-based therapies (Fig. 4.1).

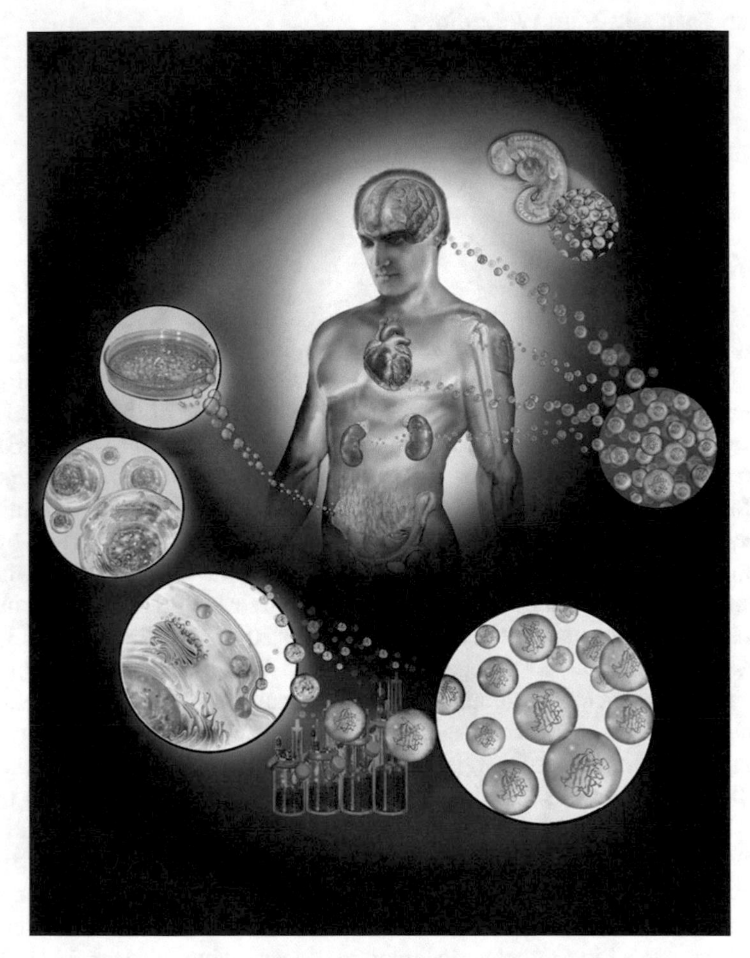

Fig. 4.1 Exosomes for repair and regeneration. Regeneration leverages mechanisms of organogenesis. Exosome mediated morphogen gradients are one such mechanism of action active in the developing embryo. Although exosomes may be isolated from any cell type or bodily secretion, in this example, exosomes are being sourced from MSC-like cell populations derived from adipose or bone marrow (pelvis). Manufacture of a clinically relevant dose will involve cell expansion in bioreactors and may include tuning or modulation of specific exosome sub-populations carrying defined payloads. Importantly, exosomes may be sourced allogeneically as a storable, "off-the-shelf" product that can be delivered to a broad patient population. Examples of organs potentially treatable with exosome-based therapeutics as suggested by preclinical data include the brain, heart, kidneys, tendon, and skin

Exosome Nomenclature

Since the first meeting of the International Society for Extracellular Vesicles (ISEV) in 2013, there have been many terms and names for secreted vesicles based primarily on their different physiological functions. This being said, "exosomes" and "microvesicles" have been widely applied and are more generic terms do designate secreted vesicles. To unify vesicle nomenclature, ISEV recommends using the term "extracellular vesicles" (EVs) as a generic term for all secreted vesicles. EVs include nanometer-scale vesicles (exosomes) as well as larger-scale nanometer vesicles (microvesicles). For the purpose of therapeutic product development, it has been determined that only exosomes and larger vesicles have potential applications owing to their relative stability (Wang et al. 2017). Although no rigorous and universally accepted definition has yet been established (Lötvall et al. 2014; Lener et al. 2015; Gould and Raposo 2013), the term *exosome* is generally understood to reference a specific class of lipid-membrane bound extracellular vesicle (EV) characterized by a diameter of 40–150 nm and density of 1.09–1.18 g/ml (Fig. 4.2a). This variability is a reflection of the isolation and purification methods used (Lane et al. 2015), and the density gradient material (e.g., sucrose or Opti-Prep) used for analysis. Indeed, some reviews cite exosomes as being 30–100 nM in size (Rashed et al. 2017). Microvesicles are larger than exosomes, and are often described as being 100–300 nM in size. The degree of overlap in the sizes for these classes of EVs also varies depending upon the technology used to make the measurement. While both a position statement (Lötvall et al. 2014) and position paper (Lener et al. 2015) have been published by key opinion leaders within the International Society of Extracellular Vesicles (ISEV) in late 2014 and 2015, a uniform consensus on how best to isolate, size, and characterize exosomes is still upcoming. It is anticipated that as the field moves forward, particularly in the area of biological function, techniques to best isolate, size, and characterize particles will be driven by which method gives the desired biological effect.

Exosome Biology

First observed by Pan and Johnstone in the late 1980s while studying the maturation process of reticulocytes into erythrocytes, it was believed that these vesicles were simply removing unnecessary proteins and other molecules from the releasing cells (Pan et al. 1985; Johnstone et al. 1987; Johnstone 1992). Formed through internal budding of the plasma membrane as multivesicular bodies within the endosome compartment, exosomes and their contents are secreted into the extracellular milieu through fusion of the endosome with the plasma membrane. Although active in a range of cellular processes and isolatable from numerous body fluids, from a regenerative medicine perspective, the critical role of exosomes is in cell-cell communication via transport of protein, mRNAs, and micro-RNAs (Valadi et al. 2007).

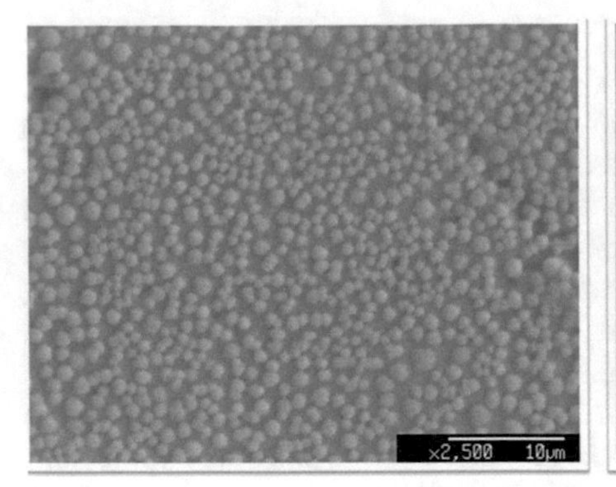

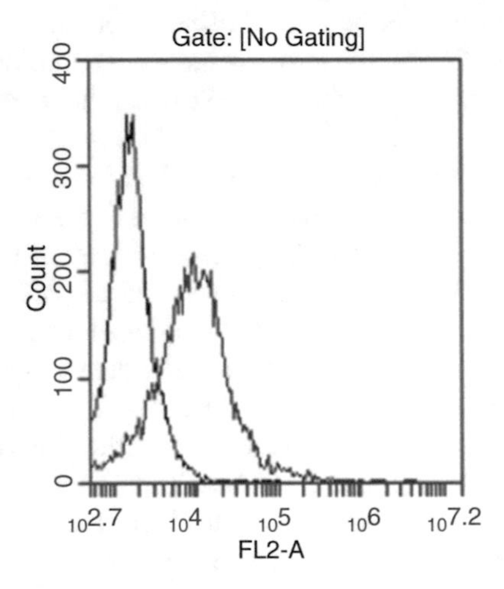

Fig. 4.2 (**a**) Scanning electron micrograph of exosomes isolated from human placental sourced MSC (2500×, left panel; 10,000×, right panel). Samples were also analyzed with a scanning electron microscope (SEM). The SEM image relies on surface and has a great depth of field, so it can produce images that are good representations of the three-dimensional shape of the sample. Another advantage of SEM is that it can produce images of sufficient quality and resolution with the samples being wet or contained in low vacuum or gas. This greatly facilitates imaging biological samples (i.e. exosomes) that are unstable in the high vacuum of conventional electron microscopes. The samples were lyophilized on the sample holder and stained using platinum (Pt) to give contrast, since many biological materials are nearly "transparent" to electrons. The Pt coating will contribute no more than 10 nm to the size of the images. Electron micrographs support the >200 nm size values obtained using the qNano. (**b**) Flow cytometry data showing exosome-mediated dye transfer to human endothelial cells. In this scatter plot diagram, the black peak represents fluorescence from the negative control population, while the red peak is fluorescence from the Dil-labeled population. The red peak is right-shifted due to incorporation of fluorescently labelled exosomes. Seventy-five percent of cells are labelled

Although such signaling generally occurs within an organism, the nematode parasite *Heligmosmoides* has been demonstrated to manipulate the innate immune response of its mouse host via secretion of exosomal elements, thereby establishing exosomes as a mechanism for interspecies transfer of RNA (Buck et al. 2014). Key protein markers that have been associated with exosomes include CD9, CD63, CD81, HSP70, HSP90, actin, and annexin (http://exocarta.org/exosome_markers; http://microvesicles.org/index.html).

Selection and Delivery of Cargo

Exosomes also contain the protein TSG101, a component of the endosomal sorting complexes required for transport (ESCRT)-I, which regulates vesicular trafficking processes. The ESCRT machinery is made up of several cytosolic protein complexes, known as ESCRT-0, ESCRT-I, ESCRT-II, and ESCRT-III. Together with a number of accessory proteins, these ESCRT complexes enable a unique mode of membrane remodeling that results in membrane bending and budding away from the cytoplasm. In this regard, TSG101 binds to ubiquitinated cargo proteins and is required for the sorting of endocytic, ubiquitinated cargos into multivesicular bodies (MVB). ECSRT complex mediated selection is not the only mechanism for getting protein cargo into exosomes: proteins may also be recruited into exosomes by virtue of their association with chaperones such as HSP70 and HSP90 (Buschow et al. 2010).

The precise mechanism by which RNA species are selected for recruitment into exosomes is less well defined. One possible mechanism involves specific sequence motifs that may function as *cis*-acting elements for targeting RNAs to EV (Batagov et al. 2011). The discovery that ESCRT-II is an RNA binding complex (Irion and St Johnston 2007) suggests that it may also function to select RNA for incorporation into EVs. Finally, the observations that MVB are sites of miRNA-loaded RISC (RNA-induced silencing complex) accumulation (Gibbings et al. 2009) and that exosome-like vesicles are considerably enriched in GW182 and AGO2 implicate functional roles of these proteins in RNA sorting to exosomes.

In order for exosomes to transfer their cargo (nucleic acid, protein), they must somehow be delivered into the targeted cell. Fluorescently labeled lipophilic dye transferred from exosomes and incorporated into cultured cells has been used to demonstrate that exosomes act to transport cargo into cells (Deregibus et al. 2007). A fusion event between exosomes and the cell membrane will transfer fluorescence from the labeled exosomes to the targeted cells (see Fig. 4.2b). These experiments identify exosomes as vectors for information transfer between cells, and highlight a specific mechanism by which one cell population may manipulate another. To this end, stable modification of cell fate by exosomes has been observed in rodent models, where lung-derived exosomes and microvesicles were shown to reprogram bone marrow cells toward a pulmonary phenotype in vitro and in vivo (Aliotta et al. 2012). Similar effects have been noted with exosomes sourced from liver

(Quesenberry et al. 2014). In addition, exosomes sourced from cancer cells can catalyze the development of tumorigenesis (Melo et al. 2014; Zhang and Grizzle 2014). Finally, in the brain, exosome mediated transfer of toxic protein aggregates such as amyloid-β and prions may represent an important mechanism for the onset of dementias and other related pathologies (Gupta and Pulliam 2014).

Exosomes: The New Paradigm in Stem Cell Biology

Multiple studies have provided evidence that the cell per se is ultimately superfluous in catalyzing observed regenerative bioactivity from cell-sourced therapeutic product candidates. Exosomes may represent an ideal, generally noncytotoxic and well-tolerated "off-the-shelf" regenerative therapy, delivering most of the potential of cell-based therapies while facilitating extensive simplification of process development and manufacturing. In a systematic review of the literature presenting preclinical animal data on the therapeutic potential of MSC-derived microvesicles including exosomes, all 13 reported studies demonstrated that treatment improved at least one clinically relevant parameter associated with organ functionality (Akyurekli et al. 2015). As an example, MSC-derived conditioned media was shown to significantly improve multiple biomarkers of renal pathophysiology in rodent models of chronic kidney disease (van Koppen et al. 2012). Mechanistically, MSC generally do not repair organ defects by differentiating into the desired tissue type, but rather function more in a regulatory role. This paradigm shift followed the demonstration that MSC can inhibit apoptosis, stimulate angiogenesis, promote endogenous cell proliferation, and interfere with the onset and progression of inflammation during tissue regeneration (Ratajczak et al. 2012). Such bioactivity is mediated through growth factor and cytokine secretion (paracrine effects) in addition to cell-cell interactions. Functional characterization of exosomes from these stem cells is a novel area of study; many of the regenerative properties previously credited to stem cells are being shown to be mediated through secreted exosomes. If valid, innovative approaches to wound healing, tissue engineering, and regenerative medicine, whereby live cell therapies could be replaced with exosomes as an active biologic, may be facilitated. The regenerative potential of exosomes may be modulated or tuned by prior exposure of the source cell population to external stimuli—for example, inflammatory conditioning of human umbilical cord blood derived MSC with IFN-γ results in MSC less able than unconditioned MSC to protect against acute ischemic renal injury in vivo (Kilpinen et al. 2013). Additional methodologies for exosome tuning may incorporate defined cell–biomaterial and 3D cell–cell interactions to regulate and further control cargo loading and exosome biogenesis (Lamichhane et al. 2015).

CD34+ cell populations have been established to induce neovascularization in preclinical studies and Phase I/II clinical trials. In in vitro and in vivo functional bioassays of angiogenesis, exosomes derived from CD34+ cells were demonstrated to mimic the effects of cells themselves, in some instances with improved potency compared to cells (Sahoo et al. 2011). Furthermore, treatment with MSC-sourced exosomes was demonstrated to block the activation of hypoxic signaling that induces pulmonary inflammation and onset of pulmonary hypertension in rodent models (Lee et al. 2012). In skin, iPSC-MSC sourced exosomes, upon injection in and around the wound bed of rodent skin wounds, were shown to significantly accelerate wound healing, collagen synthesis and revascularization of the wound site (Zhang et al. 2015a); see illustrative example Fig. 4.3. In proof of concept studies of myocardial infarct in the rat, exosomes derived from cardiosphere-derived cells (CDC), were demonstrated to enhance cardiac functionality, decrease scar mass, increase viable tissue mass and infarct wall thickness relative to exosomes sourced from dermal fibroblasts or media controls. Importantly, injection of CDC-exosomes at 21 days post-infarct, a time-point with a well-established scarification profile, resulted in significant growth of new myocardial tissue as well as functional improvements consistent with a true regenerative outcome. miRNA profiling of CDC-sourced and fibroblast-sourced exosomes identified *mir-146a* as a potential active biological ingredient catalyzing the observed functional activity; aspects of CDC-exosome bioactivity could be reproduced by treatment with *mir-146a* (Ibrahim et al. 2014).

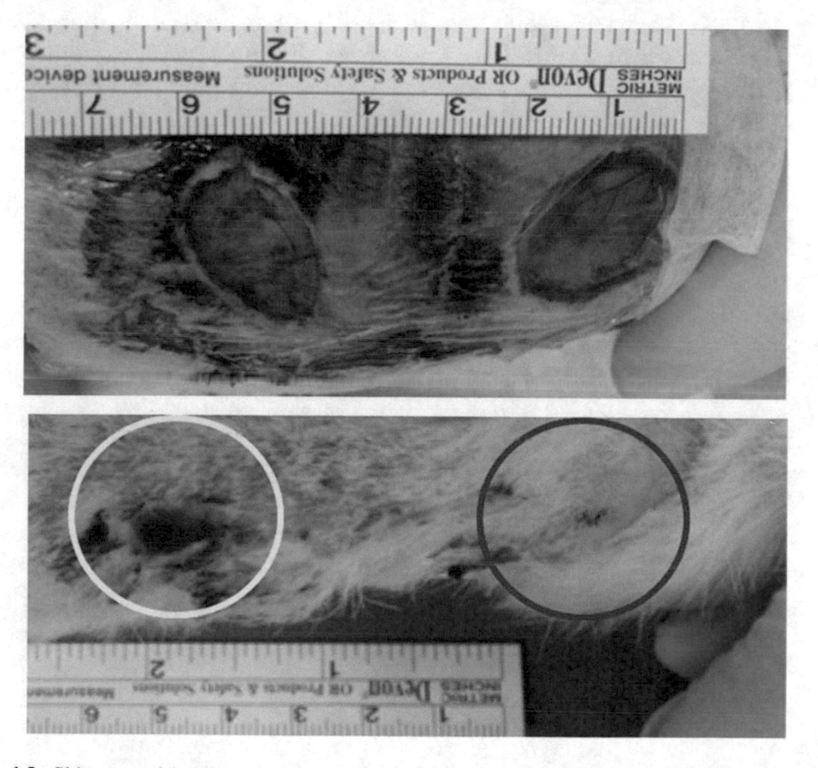

Fig. 4.3 Skin wound healing model in rodent. Illustrative example. Top panel: 2 cm diameter complete removal of dermis from dorsal surface of rodent. One wound treated with MSC-sourced exosomes, other treated with saline as control. Bottom panel: 2 weeks post-injury, both wounds have healed substantially, but wound treated with MSC-sourced exosomes (red circle) has healed substantially faster

Box 4.1 Regeneration Leverages Developmental Signaling Mechanisms: Exosome-Mediated Transfer of Morphogens

Organ regeneration technologies aim to restore the original structure and functionality of a diseased organ. In general, healing responses within mammals are characterized by fibrosis and scar tissue formation, not regeneration. Nevertheless, developing mammalian fetuses during the first trimester will typically present wound healing without fibrosis and scar tissue formation (Adzick and Lorenz 1994). Additionally, compensatory hyperplasia of mammalian kidney or liver secondary to partial nephrectomy or hepatectomy, remodeling of epidermis or bone consequent to injury and regeneration of limb digit tips in humans and mice post-amputation are all examples of regenerative outcomes in adult mammals indicative of an innate regenerative potential within adult mammals (reviewed by Roy and Gatien 2008).

However, model organisms such as *Hydra*, planaria, zebrafish, *Xenopus* and urodeles (salamanders) present the clearest examples of regenerative outcomes secondary to injury. In these systems, cell-based strategies harnessing pluripotent and tissue specific stem cells as well as dedifferentiation have been leveraged to mediate the regeneration of whole limbs and organs (Tanaka and Reddien 2011). Systematic experimentation with limb regeneration in urodeles has permitted the decipherment of key mechanistic pathways of regeneration at the molecular level. Activation of salient signaling cascades including p53, TGF-β, Delta, ppRB and Wnt/β-catenin have all been associated with limb regeneration (Roy and Gatien 2008). These signaling pathways catalyze a sequence of instructive interactions between mesodermal and ectodermal cell populations that are ultimately responsible for lineage specification (Wessels 1977). In addition, the methodical depletion of macrophages within the first 24 h subsequent to limb amputation in urodeles has been demonstrated to lead to permanent failure of limb regeneration, extensive fibrosis, and disregulation of transcriptional patterns associated with synthesis of extracellular matrix (ECM) components (Godwin et al. 2013). This specific sequence of cellular events associated with regeneration of the urodele limb as defined below recapitulates aspects of embryonic organogenesis and may serve as a model system for establishing the existence of similar pathways in mammals (Buckley et al. 2012).

1. The open wound is enclosed by wound epithelium to form a permissive epithelial structure referred to as the apical ectodermal cap (AEC).
2. Upregulation of matrix metalloproteinase (MMP) expression catalyzes structural reorganization of the ECM.
3. Dedifferentiation of cell populations takes place proximal to the plane of amputation.
4. Proliferation and migration of dedifferentiated cells is observed under the AEC.

5. Induction of a blastema, a mass of mesenchymal cells that will eventually redifferentiate to create the new muscles, bones, nerves, and tendons required to regenerate a functional limb.

An understanding of these stages has already been applied to accelerate regenerative outcomes in mammals. For example, application of MMP1 to digit remnants of adult mice with amputation at the mid-second phalanx significantly improved regeneration of soft tissue and observed rates of wound closure. More multipotent progenitor cells, capillary vasculature and neuromuscular related tissues were also noted (Mu et al. 2013). Furthermore, recent data on regenerative outcomes in mammals from tissue engineering of bladder, esophagus and intestine provides additional evidence of the existence of a regenerative pathway in adult mammals mimicking aspects of that observed in urodeles, including formation of a neoblastema (Basu 2014; Basu and Ludlow 2010, 2011, 2012b; Basu et al. 2011a, b, 2012a, b, 2013; Basu and Bertram 2014). This regenerative pathway is characterized by a dependence on adequate vascularization and innervation at the site of regeneration. Importantly, these observations provide insight into a potential mechanism of action for cell-sourced therapies characterized broadly as *instructive* signaling between mesenchymal cells or cell-derived by products of the regenerative implant and host epithelial cell populations. This insight may be harnessed to facilitate development of novel regenerative products.

The mechanistic link between developmental and regenerative biology predicts that potential regenerative therapies may leverage or manipulate the fundamental signaling pathways governing cellular self-organization during embryonic organogenesis. For example, the highly convoluted nature of developing epithelia mandates the existence of an efficient mechanism for morphogen transport across the plasma membrane to establish the short and long range morphogen gradients central to assembly of the developing embryo. To this end, the observation that morphogens including *Wingless* and *Hedgehog* are closely associated with the plasma membrane, as opposed to freely diffusing across the cytosol, strongly suggests the existence of a membrane-based transcytotic vesicular mechanism for establishment of the morphogen gradient. Evidence from the developing *Drosophila* embryo demonstrates that the establishment of gradients of the morphogen *Wingless* during pattern formation of the imaginal disc epithelium occurs at least in part through membrane bound exosome-like particles called "argosomes" (Greco et al. 2001). In *C. elegans*, an apical secretion pathway mediated by the membrane bound V0 sector of the vacuolar H+-ATPase controls secretion of Hedgehog-like proteins within exosomes (Liegeouis et al. 2006). Finally, the specification of left/right asymmetry in the developing mouse requires the

(continued)

Box 4.1 (continued)
exosome-mediated transport of the morphogens *Sonic Hedgehog* and retinoic acid in response to FGF-signaling (Tanaka et al. 2005). Vertebrate *Sonic Hedgehog* has been reported to be secreted within two overlapping populations of exosomes, presenting distinctive accessory signaling proteins. Coexpression of integrins was required together with *Sonic Hedgehog* to activate certain *Sonic Hedgehog* target genes during differentiation of mouse ES cells, suggesting the existence of a mechanism for fine-tuning exosome-based morphogen gradients by presentation of distinctive subcategories of morphogen presenting exosomes (Vyas et al. 2014). Finally, the *Xenopus* cleavage stage blastocoel is bridged by multiple arrays of parallel filopodia, that facilitate direct interaction between nonadjacent blastomeres; these filopodia in turn fragment into microvesicles (including exosomes) whose subsequent resorption identifies a specific mechanism for the potential transfer of morphogens across the developing embryo (Danilchik et al. 2013).

In rodent models of stroke, intravenous administration of MSC-sourced exosomes was demonstrated to enhance functional recovery while accelerating neurite remodeling, neurogenesis and angiogenesis (Xin et al. 2013; Zhang et al. 2015b). Healing and repair of the tendon is another objective for intervention with cell-based therapeutics. The direct incorporation of MSC into specialized sutures used to repair defined tendon injury was demonstrated to significantly accelerate healing as assessed by histology or biomechanical loading (Adams et al. 2014). Similar outcomes were observed in studies evaluating the impact of direct microinjection of human-sourced MSC into the injury site of a rodent presenting collagenase-induced Achilles tendon injury (Machova Urdzikova et al. 2014). Although not causally attributable to exosomes, these results in the context of the broader paracrine-based model for MSC mechanism of action imply that tendon injuries may indeed be candidates for treatment with exosome-based therapies. As considerable effort has been applied toward the application of MSC for repair of tendon injuries in the horse, any development of an "off-the-shelf" exosome-based therapy represents a significant value proposition for the veterinary market (Tetta et al. 2012).

Exosomes as a New Disruptive and Safer Therapeutic for Regenerative Medicine

The application of naturally occurring secreted vesicles such as exosomes might allow overcoming toxicity or immunogenicity associated with other developed carrying agents like liposomes or nanoparticles (Fleury et al. 2014). Relative to cells, exosomes are more stable and storable, have no risk of aneuploidy, a lower possibility of immune rejection following in vivo allogeneic administration, and may provide an alternative therapy for various diseases (Yu et al. 2014). In short, exosomes do not elicit an acute immune rejection response in the recipient like cells do. As such, their utility as a therapeutic is vastly expanded over the use of cells since exosomes may be applied allogeneically. Exosomes also present less of a health and safety risk for such adverse events as tumor or emboli formation, which are often major concerns for cellular therapies, since exosomes are both nonviable and much smaller in size compared to live cells. Interestingly, exosomes derived from adipose-sourced MSC taken from cancer patients presented comparable miRNA expression profiles to exosomes sourced from noncancerous controls (Garcia-Contreras et al. 2014), suggesting that, at least for these patients, exosomes may not mediate the de novo induction of cancer. For companies' currently developing exosomes as a therapeutic, it is anticipated that manufacturing and storage of these nonviable yet biologically active microvesicles is expected to be far less complicated and costly compared to cells (Box 4.1).

Exosomes as Vectors for Repair: Skin

The skin is frequently injured by acute and chronic wounds, such as diabetic skin ulcerations or extensive burns. In a recent study, exosomes from hiPSC-MSC were found to exert beneficial effects on granulation tissue formation and angiogenesis, which are two critical phases of the wound-healing process (Zhang et al. 2015a). In addition, exosomes from these cells facilitated a significant therapeutic effect during cutaneous wound healing, supporting the notion that exosomes may be used as therapeutic tools in wound healing. Mechanistically, Wnt4 delivered by exosomes appears to be the key mediator in this type of skin healing and repair (Zhang et al. 2014), see illustrative example Fig. 4.3. Keloids represent the most extreme example of cutaneous scarring as a pathological response to wound healing. Enhanced STAT3 expression and phosphorylation has been observed in keloid scar tissue and in cultured keloid fibroblasts (Lim et al. 2006). This type of scarring has an overabundance of collagen deposition, contributing to its lack of softening, flattening, and remodeling over time. In vitro inhibition of STAT3 phosphorylation has been shown to contribute to the loss of collagen production in these cells. This raises the interesting possibility that inhibitors of STAT3 phosphorylation may be useful in prospectively treating burn wounds in vivo to reduce keloid scar formation. Indeed, treatment with mouse exosomes or exosomes derived from MSCs isolated from human umbilical cord stroma completely abrogated STAT3 phosphorylation due to hypoxia (Lee et al. 2012).

In diabetic rat wound healing models, treatment with MSC conditioned media improved wound closure rates, collagen production, and angiogenesis (Li et al. 2015; Wang et al. 2012). In a recent study, EVs from hiPSC-MSC were found to exert beneficial effects on granulation tissue formation and angiogenesis (Sen et al. 2009), which are two critical phases of the wound-healing process. In addition, EVs from these cells facilitated a significant therapeutic effect during cutaneous wound healing, supporting the hypothesis that EVs may be used as therapeutic tools. Indeed, Hu et al. (2016) recently demonstrated wound healing properties of exosomes, a subset of EVs, isolated from human adipose mesenchymal stem cells following their localized and intravenous injection.

Exosomes as Cosmeceuticals; Vectors of Rejuvenation

Cosmeceuticals are cosmetic products with biologically active ingredients purporting to have medical or drug-like benefits. The "cosmeceutical" label applies only to products applied topically, such as creams, lotions and ointments. Liposomes are well-known vesicular cosmetic delivery systems (Madsen and Andersen 2010). For example, liposomes may potentially be used to deliver avobenzone (a sunscreen) and arbutin (a skin whitening agent) in a differential manner such that the sunscreen is retained at the skin surface while the whitening agent is delivered further into the dermal strata (Liu et al. 2013). Nebulized liposomes have also been evaluated for delivery of vitamin K1 into the skin (Campani et al. 2014). Their topical application offers several advantages including increased moisturization, restoring action, biodegradability, biocompatibility, and extended and slow dermal release. Their similar structure to biological membranes allows for penetration into the epidermal barrier (Rahimpour and Hamishehkar 2012).

There are already cosmetic products on the market which contain stem cell conditioned medium (http://www.lifelineskincare.com/skin-care-science), and fibroblast cell conditioned medium (https://www.truthinaging.com/review/regenica-advanced-rejuvenation-overnight-repair). This medium contains growth factors, EVs, and potentially cell waste products. Given the structural similarities between liposomes and exosomes, along with the inclusion of conditioned cell medium in selected skin care products, it seems reasonable to expect that exosomes will soon find their place in the cosmeceutical industry much like liposomes and other cell-derived products have (see Box 4.2). If exosome turn out to be the active ingredient in skin products that contain conditioned cell medium, isolation of this active for product inclusion may enhance the product's effects. Indeed, a skin care product was launched in September of 2016 which incorporates 150 million exosomes, and is reported to reduce wrinkles, increase skin luminescence, and accelerate cell renewal (https://exoskinsimple.com/products/bio-digital-perfection-moisturizer).

Box 4.2 Cell Secretome as Cosmeceutical: A Potential Role for Exosomes

Although data demonstrating the impact of stem cell or cell-sourced products in modulating the biologic and biophysical properties of aging skin does not in itself prove that exosomes may be useful or relevant in these applications, such a role may reasonably be extrapolated based on regenerative outcomes associated with exosomes in other systems. To this end, cell-derived secretomic extracts have demonstrated value as cosmeceuticals for rejuvenation of aging skin as well as for the promotion of hair growth. Evidence for a direct impact of stem cell-derived secretomic factors in promoting skin rejuvenation is provided by randomized, investigator blinded "split-face" studies where cell derived secretomic extracts are delivered by microneedle to one half of a subject's face. A control, mock procedure using just the microneedle is applied to the other half. In such studies, a statistically significant improvement in skin pigmentation, wrinkling and roughness was noted in the presence of the cell-derived secretome (Seo et al. 2013; Lee et al. 2014). Similarly, the intradermal injection of GCSF-mobilized PBSCs from young pig could rejuvenate cheek skin of aged pigs as shown by increased levels of collagen, elastin, hyaluronic acid and CD44, involucrin, integrin as well as increases in proliferative capacity in the basal layer (Harn et al. 2013). In mouse model of wrinkling created by UV-B irradiation of hairless mice, wrinkling, dermal thickness and collagen content were all improved by injection of adipose-derived stem cells. In vitro studies implicated secretomic factors sourced from the adipose stem cells as potentially important in mediating their antiaging properties (Kim et al. 2009). Conditioned media derived from human dermal fibroblasts were shown to ameliorate the UV-A induced upregulation of MMP1 and associated downregulation of collagen and TIMP1 transcripts as well as promoting migration and inhibiting apoptosis in vitro (Shim et al. 2013). Regenerative cycling of hair waves has been studied in mouse skin. In this model, such cycling slows down with increasing age; however, this behavior is non-cell autonomous, such that transplantation of aged mouse skin into a young host rescues regenerative cycling, thus implicating secretomic factors as inductive for hair follicle regeneration (Chen et al. 2014). Conditioned media from adipose stem cells, upon intradermal injection into alopecia patients using a "split-scalp" study design, has been reported to significantly promote hair growth (Fukuoka and Suga 2015). Finally, the observation that miR-214, acting through the Wnt pathway, regulates both skin morphogenesis and hair follicle development opens the possibility of leveraging discrete populations of defined microRNAs delivered through exosomes for triggering regeneration of hair (Ahmed et al. 2014).

Scalable Production of Exosomes

Preclinically, the use of MSC-derived exosomes is strongly associated with improved organ function following injury and may be useful for inhibiting tumor growth (Akyurekli et al. 2015). Exosomes have already been tested as a cancer vaccine in the clinic (Escudier et al. 2005; Morse et al. 2005; Dai et al. 2008). These studies were limited to particles produced during short-term ex vivo culture of autologous dendritic cells. While limited in scope, this work is significant because the exosomes were deemed safe in the small clinical trials conducted (Escudier et al. 2005). As with any biologic, scalable production of the active ingredient must be achieved to have relevance as a readily available and commercially feasible therapeutic. Unfortunately, the process by which these exosomes were manufactured for these studies provides little guidance for large-scale cGMP manufacturing of exosomes needed for more comprehensive clinical trials. In addition, hundreds of micrograms to milligram quantities of exosomes may be needed to treat many patients in a clinical trial. Senescence of the cells from which exosomes are being manufactured represents an intrinsic limitation on final absolute amounts. Loss of actively growing cells will most certainly effect exosome production, which in turn would jeopardize trial outcomes. One approach to address the growth arrest/senescence issue is cell immortalization. Indeed, *MYC* transformation may represent a practical strategy in ensuring an infinite supply of cells for production of exosomes in the milligram range as a therapeutic agent (Chen et al. 2011). In addition, the increased proliferative rate of cells should reduce time for cell production, thus reducing production costs.

Another hurdle to overcome is how to culture a sufficient number of cells to produce enough conditioned medium from which milligram quantities of exosomes may be isolated. From a cGMP standpoint, cell culturing in a closed system is preferred. One approach may be the use of hollow-fiber cell bioreactors, as a cGMP-compliant closed culture system, for culturing large numbers of cells to produce large quantities of exosomes. A bioreactor approach should also abolish the need to continually passage cells during a production run, alleviating the need for huge numbers of plastic tissue culture vessels while reducing medium volume. In the long term, use of bioreactors has the potential to increase efficiency of exosome production while simultaneously reducing cost-of-goods. A white paper describing the culture of placental derived MSCs in a hollow fiber bioreactor is a useful guide for starting to address the scalable production of exosomes (Cadwell n.d.). Preliminary results have shown the bioreactor yield is in milligrams, approximately tenfold greater than cultures grown in T-flasks and cell factories, while simultaneously resulting in a higher concentration/ml conditioned medium (Basu et al. 2015). Alternatively, a device that uses centrifugal force and a filter with microsized pores has been used to generate large quantities of cell-derived nanovesicles (Jo et al. 2014). The nanovesicles produced are similar in size and membrane structure to exosomes, and they contain intracellular RNAs ranging from microRNA to mRNA, intracellular proteins, and plasma membrane proteins. The quantity of nanovesicles produced using the device is 250 times the quantity of naturally secreted exosomes, the quantity of intracellular contents in nanovesicles is twice that in exosomes, and these particles can transfer RNAs to target cells. Side-by-side studies between these manufactured nanovesicles and naturally produced exosomes will determine the utility of this device in producing large quantities of a therapeutically relevant biologic. Finally, an additional factor for consideration is that any therapeutically relevant bioactivity may be a function of an exosome-mediated secretory milieu that is by definition heterogeneous and not necessarily associated with any single molecule or medicinal agent. As a precedent, a heterogeneous population of renal cells has been developed as a cell-based therapeutic for chronic kidney disease—no single, definable cell population is understood to mediate observed regenerative outcomes (Basu and Ludlow 2012a). It is likely that exosome-based therapeutics and cosmeceutics catalyze their bioactivity as a function of their difficult-to-define heterogeneous nature as admixtures of medicinal agents.

Manufacturing

Gimona et al. (2017) have just published an excellent Opinion Paper on Manufacturing of Human Extracellular Vesicle-Based Therapeutics for Clinical Use. Several of their key points to consider are excerpted below:

1. Is the therapeutic being developed for a small patient population or for a large number of potential patients? This is an important decision, as it affects the amounts of EVs to be manufactured and for the general question of scalability. This decision also affects the design and amount of nonclinical (in vitro and

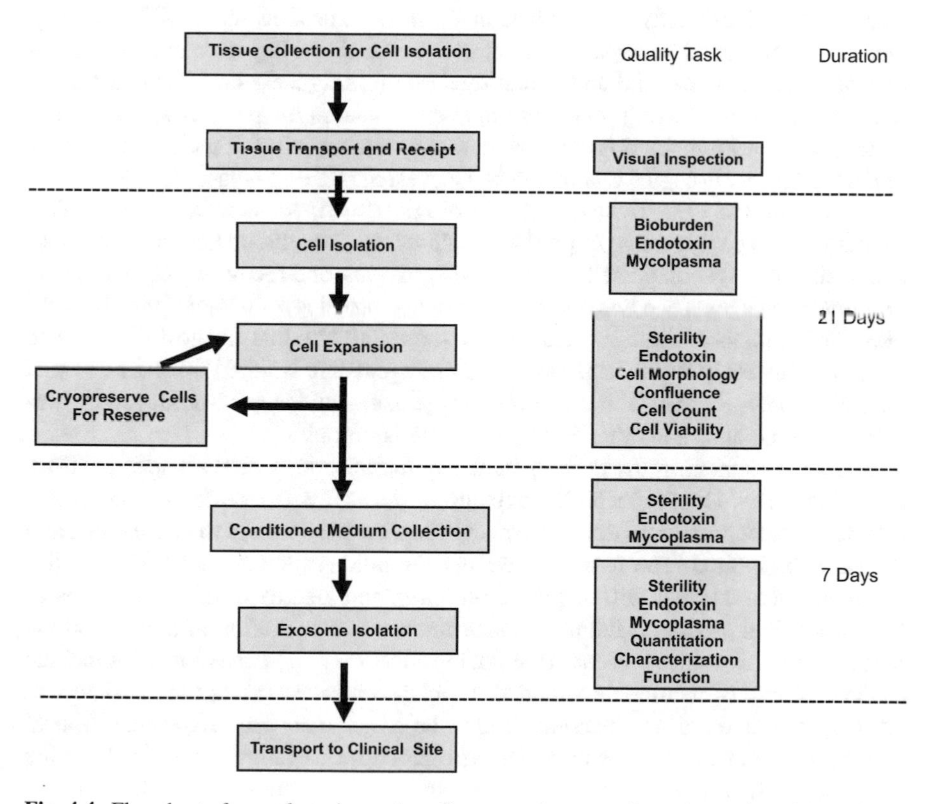

Fig. 4.4 Flowchart of manufacturing and quality-control strategy for exosome-based therapies

in vivo animal) data and clinical testing of the future biological drug that is needed.
2. Adequate therapeutic batch sizes must be planned and evaluated during process and product development.
3. Will the product be designed primarily either to address a clear unmet medical need or to compete against multiple existing treatment options?
4. Route of application, which is dependent upon the disease to be treated, needs to be defined early during product development.
5. Assuming that the therapeutic is based on isolation for human cells, either an allogeneic or autologous use has to be evaluated in a risk-based approach (Fig. 4.4).

Regulatory Requirements for Manufacturing and Quality Control

The regulatory requirements placed upon the biotechnology industry for production of medicinal products are quite demanding. Manufacturing of exosomes for therapeutic applications needs to take place in a tightly controlled and qualified setting. Quality systems must be in place to control the manufacturing environment, validation of equipment, material and operational controls. Process controls and validation are critical to meeting regulatory agency standards for product approval. For therapeutic development, it is anticipated that exosomes will fall under the purview of the Center for Biologics Evaluation and Research (CBER)—vaccines, blood, and biologics—of the FDA (www.fda.gov/BiologicsBloodVaccines/default.htm) This Center reviews a wide range of products such as vaccines, blood and blood components, allergenics, somatic cells, gene therapy, tissues, and recombinant therapeutic proteins. Such agents can be composed of sugars, proteins, or nucleic acids or complex combinations of these substances (exosomes fall into this category), or may be living entities such as cells and tissues. These agents are isolated from a variety of natural sources—human, animal, or microorganism—and may be produced by biotechnology methods and other cutting-edge methodologies.

A crucial element in the development any therapeutic is product specifications related to purity, identity, quantity, potency, and sterility. If release criteria are not met due to deviations from defined expectations, the product has to be rejected, and its use is prohibited. The more precise the definitions for EV-based therapeutics become, more emphasis will be placed on purity and identity of the batch preparation. If the final product is defined as exosomes, a demonstration of purity, and the percentage of exosomes present in the final product may be in question. EVs include a broad variety of membrane-bounded vesicles, while exosomes are restricted, at least, by size and surface markers. It may be anticipated that large-scale clinical manufacturing of exosomes will be less stringent on segregating other EVs from the preparation. Such a scenario my result in an increasing number of batches being rejected. To alleviate this, perhaps the heterogeneity of secretome-based preparations need to be acknowledged and accepted, and in so doing a terminology that embraces all biological components and therapeutic aspects without eliminating the central claim needs to be used. It may be more appropriate to consider the resulting product to be a vesicular secretome fraction (Wang et al. 2017). In the event that the identity and purity of a biologic therapeutic cannot be better defined, a common principle for such early development of biologicals anticipates that "the process is the product," and EV therapeutics certainly fall into this category.

Below is a potential example, based on our experiences in developing several cell therapeutic and tissue engineered products, of a flow-diagram for development of exosomes as a therapeutic illustrating what FDA might look for in a manufacturing scheme. At left—cells are isolated, cultured, expanded, and exosomes isolated from conditioned medium. This schematic assumes that cells will be extracted from a specific tissue type for use in exosome isolation; for cells already isolated, the steps will begin at the cell expansion stage. The quality tasks, which FDA is most interested in during the manufacturing process, are in boxes at right. Notice that they are heavily focused on testing for contamination by microorganisms, cell number, and cell viability during multiple steps of the process. Testing of the final product, the exosomes, also includes testing for microorganism contamination. In addition, the exosomes must be characterized, which will include the determination of physicochemical properties, biological activity, immunochemical properties, purity, and impurities. This is necessary to establish the safety and efficacy profile of the product.

Synthetic Exosomes and Exosome Mimetics

Exosomes by their nature represent a heterogeneous, incompletely characterized biologic product. In addition, it remains to be established whether comparable lots of exosome preparations are routinely and consistently isolatable at large scale. Together with the somewhat tedious and time-consuming nature of the exosome isolation and manufacturing process (Marcus and Leonard 2013), see also above, these factors have triggered attempts to design and synthesize exosome-like particles or exosome mimetics that could potentially be made at much larger scale. Such particles have a potentially significant advantage in being fully definable at the lipidomic, proteomic, and transcriptomic levels (Kooijmans et al. 2012). For example, ES cell-derived nanovesicles that mimic exosomes have been created by extruding living ES cells through microfilters and shown to promote proliferation of primary murine skin fibroblasts (Jeong et al. 2014). Other methodologies currently under development include exosomes as vectors for microRNAs, siRNAs, or other defined protein cargo (Marcus and Leonard 2013).

Exosomes as Biomarkers for Disease and Regeneration

Finally, the presence of exosomes in multiple body secretions and fluids may be leveraged as a mechanism to monitor disease phenotypes or regenerative outcomes associated with a therapy. For example, the presence of certain microRNA biomarkers in urine sourced exosomes may be leveraged to evaluate development of renal fibrosis; conversely, the presence of exosomes expressing CD133 or other stem and progenitor cell proteins may be an indicator of regenerative activity within the kidney. Although molecular assays have been proposed to facilitate the rapid assessment of renal regeneration associated with application of cell-based therapies (Genheimer et al. 2012), the ability to monitor such outcomes merely by measurement of certain defined urinary exosomes would represent a significant improvement (Ranghino et al. 2015).

Concluding Remarks and Future Perspectives

As evidence from multiple experimental systems accumulates implicating the secretome in general and vesicular components such as exosomes in particular as principal mechanistic agents catalyzing the observed regenerative bioactivity of cell-based products, the parallel emphasis on product development is transitioning to increasingly focus on exosome-based therapies over cell-based therapies. Exosomes present considerable advantages over cells for manufacturing, storage, handling, product shelf life and their potential as a ready to go biologic. Globally, at least one clinical trial of a MSC-sourced exosomes for improvement of β-cell mass in type 1 diabetes patients has been reported (https://clinicaltrials.gov/ct2/show/NCT02138331?term=exosome&rank=4), we anticipate many more studies will be initiated in the next 1–5 years.

Box: Outstanding Questions?

Intellectual property claims surrounding exosomes and their applications for regenerative therapies remains to be clarified.

Mechanistically, which proteins or nucleic acids being transported by exosomes mediate observed regenerative outcomes? Or is regeneration a function of a heterogeneous composite of multiple bioactive exosome sub-populations?

Can the bioactivity of exosomes be recapitulated by synthetic, exosome-like particles?

Can exosomes sourced from non-stem cell populations also catalyze clinically relevant regenerative outcomes?

Glossary

Allogeneic Cells or cell-sourced materials derived from a donor source genetically dissimilar to the intended recipient are said to be allogeneic. Such biologics are typically immune-privileged.

Cosmeceutical Topically applied cosmetic products with biologically active ingredients purporting to have medical or drug-like benefits.

Exosome 40–100 nm membrane bound vesicles that mediate transfer of proteins and nucleic acids across cellular boundaries.

Keloid Scar tissue formed at site of healed skin injury composed of either type III (early phase) or type I (late phase) collagen.

Potency Defined by FDA as *the specific ability or capacity of the product, as indicated by appropriate laboratory tests or by adequately controlled clinical data obtained through the administration of the product in the manner intended, to effect a given result.* Potency is an important quality control criteria for all cell-based biologics, see Basu and Ludlow (2014) and Guthrie et al. (2013) for detailed discussion of potency assays and metrics for regenerative medicine and tissue engineered products.

References

Adams SB et al (2014) Stem cell-bearing suture improves Achilles tendon healing in a rat model. Foot Ankle Int 35:293–299

Adzick NS, Lorenz HP (1994) Cells, matrix, growth factors and the surgeon. The biology of scarless fetal wound repair. Ann Surg 200:10–18

Ahmed MI et al (2014) MicroRNA-214 controls skin and hair follicle development by modulating the activity of the Wnt pathway. J Cell Biol 207:549–567

Akyurekli C et al (2015) A systematic review of preclinical studies on the therapeutic potential of mesenchymal stromal cell-derived microvesicles. Stem Cell Rev 11:150–160

Aliotta JM et al (2012) Stable cell fate changes in marrow cells induced by lung-derived microvesicles. J Extracell Vesicles. https://doi.org/10.3402/jev.v1i0.18163

Basu J (2014) An organ regeneration platform for industrial production of hollow neo-organs, cells and biomaterials in regenerative medicine. www.intechopen.com/books/cells-and-biomaterials-in-regenerative-medicine/an-organ-regeneration-platform-for-industrial-production-of-hollow-neo-organs

Basu J, Bertram T (2014) Regenerative medicine of the gastrointestinal tract. Toxicol Pathol 42:82–90

Basu J, Ludlow JW (2010) Platform technologies for tubular organ regeneration. Trends Biotechnol 28:526–533

Basu J, Ludlow JW (2011) Tissue engineering of tubular and solid organs: an industry perspective. In: Wislet-Gendebein S (ed) Advances in regenerative medicine. Intech Open, Croatia

Basu J, Ludlow JW (2012a) Developmental engineering the kidney: leveraging principles of morphogenesis for renal regeneration. Birth Defects Res C Embryo Today 96:30–38

Basu J, Ludlow JW (2012b) Developments in tissue engineered and regenerative medicine products, a practical approach. Woodhead Publishing, Cambridge, UK

Basu J, Ludlow JW (2014) Cell-based therapeutic products: potency assay development and application. Regen Med 9:497–512
Basu J et al (2011a) Regeneration of rodent small intestine tissue following implantation of scaffolds seeded with a novel source of smooth muscle cells. Regen Med 6:721–731
Basu J et al (2011b) Functional evaluation of primary renal cell/biomaterial Neo-Kidney Augment prototypes for renal tissue engineering. Cell Transplant 20:1771–1790
Basu J et al (2012a) Regeneration of native like neo-urinary tissue from non-bladder cell sources. Tissue Eng Part A 18:1025–1034
Basu J et al (2012b) Extension of bladder based organ regeneration platform for tissue engineering of esophagus. Med Hypotheses 78:231–234
Basu J et al (2013) Tissue engineering of esophagus and small intestine in rodent injury models. Methods Mol Biol 1001:311–324
Basu J et al (2015) MSC sourced exosomes as therapeutic agents for wound healing and skin regeneration: from scaled production to functional regenerative outcomes in vitro and in vivo. International Society for Stem Cell Research Annual Meeting, Stockholm
Batagov AO et al (2011) Identification of nucleotide patterns enriched in secreted RNAs as putative cis-acting elements targeting them to exosome nano-vesicles. BMC Genomics 12(Suppl 3):S18
Buck SH et al (2014) Exosomes secreted by nematode parasites transfer small RNAs to mammalian cells and modulate innate immunity. Nat Commun 5:5488
Buckley G et al (2012) Denervation affects regenerative responses in MRL/MpJ and repair in C57BL/6 ear wounds. J Anat 220:3–12
Buschow SI et al (2010) MHC class II-associated proteins in B-cell exosomes and potential functional implications for exosome biogenesis. Immunol Cell Biol 88:851–856
Cadwell JS. Culture of placental derived cells in a hollow fiber bioreactor cartridge. http://fibercellsystems.com/documents/FibercellSystemsPlacental%20Stem%20Cell%20Culture.pdf
Campani V et al (2014) Development of a liposome based formulation for vitamin K1 nebulization on the skin. Int J Nanomedicine 9:1823–1832
Caplan AI, Correa D (2011) The MSC: an injury drugstore. Cell Stem Cell 9:11–15
Chen TS et al (2011) Enabling a robust scalable manufacturing process for therapeutic exosomes through oncogenic immortalization of human ESC-derived MSCs. J Transl Med 9:47
Chen CC et al (2014) Regenerative hair waves in aging mice and extra-follicular modulators Follistatin, Dkk1 and Sfrp4. J Invest Dermatol 134:2086–2096
Dai S et al (2008) Phase I clinical trial of autologous ascites-derived exosomes combined with GM-CSF for colorectal cancer. Mol Ther 16:782–790
Danilchik M et al (2013) Blastocoel spanning filopodia in cleavage stage Xenopus laevis: potential roles in morphogen distribution and detection. Dev Biol 382:70–81
Deregibus MC et al (2007) Endothelial progenitor cell derived microvesicles activate an angiogenic program in endothelial cells by a horizontal transfer of mRNA. Blood 110:2440–2448
Escudier B et al (2005) Vaccination of metastatic melanoma patients with autologous dendritic cell (DC) derived-exosomes: results of the first Phase I clinical trial. J Transl Med 3:10
Fleury A et al (2014) Extracellular vesicles as therapeutic tools in cardiovascular diseases. Front Immunol 5:370
Fukuoka H, Suga H (2015) Hair regeneration treatment using adipose-derived stem cell conditioned medium: follow-up with trichograms. Eplasty 15:e10
Garcia-Contreras M et al (2014) Therapeutic potential of human adipose-derived stem cells (ADSCs) from cancer patients: a pilot study. PLoS One:e113288
Genheimer G et al (2012) Molecular characterization of the regenerative response induced by intrarenal transplantation of selected renal cells in a rodent model of chronic kidney disease. Cells Tissues Organs 196:374–384
Gibbings DJ et al (2009) Multivesicular bodies associate with components of miRNA effector complexes and modulate miRNA activity. Nat Cell Biol 11:1143–1149
Gimona M et al (2017) Manufacturing of human extracellular vesicle-based therapeutics for clinical use. Int J Mol Sci 18(6). https://doi.org/10.3390/ijms18061190

Godwin JW et al (2013) Macrophages are required for adult salamander limb regeneration. Proc Natl Acad Sci U S A 110:9415–9420
Gould SJ, Raposo G (2013) As we wait: coping with an imperfect nomenclature for extracellular vesicles. J Extracell Vesicles 2. https://doi.org/10.3402/jev.v2i0.20389
Greco V et al (2001) Argosomes: a potential vehicle for the spread of morphogens through epithelia. Cell 5:633–645
Gupta A, Pulliam L (2014) Exosomes as mediators of neuroinflammation. J Neuroinflammation 11:68
Guthrie K et al (2013) Potency evaluation of tissue engineered and regenerative medicine products. Trends Biotechnol 31:505–514
Harn HJ et al (2013) Rejuvenation of aged pig facial skin by transplanting allogeneic granulocyte colony stimulating factor induced peripheral blood stem cells from a young pig. Cell Transplant 22:755–765
Hu L et al (2016) Exosomes derived from human adipose mensenchymal stem cells accelerates cutaneous wound healing via optimizing the characteristics of fibroblasts. Sci Rep 6:32993
Ibrahim AG et al (2014) Exosomes as critical agents of cardiac regeneration triggered by cell therapy. Stem Cell Rep 2:606–619
Irion U, St Johnston D (2007) bicoid RNA localization requires specific binding of an endosomal sorting complex. Nature 445:554–558
Jeong D et al (2014) Nanovesicles engineered from ES cells for enhanced cell proliferation. Biomaterials 35:9302–9310
Jo W et al (2014) Large-scale generation of cell-derived nanovesicles. Nanoscale 6:12056–12064
Johnstone RM (1992) The Jeanne Manery-Fisher Memorial Lecture 1991. Maturation of reticulocytes: formation of exosomes as a mechanism for shedding membrane proteins. Biochem Cell Biol 70:179–190
Johnstone RM et al (1987) Vesicle formation during reticulocyte maturation. Association of plasma membrane activities with released vesicles (exosomes). J Biol Chem 262:9412–9420
Justewicz DM et al (2012) Characterization of the human smooth muscle cell secretome for regenerative medicine. Tissue Eng Part C Methods 18:797–816
Kilpinen L et al (2013) Extracellular membrane vesicles from umbilical cord blood derived MSC protect against ischemic acute kidney injury, a feature that is lost after inflammatory conditioning. J Extracell Vesicles. https://doi.org/10.3402/jev.v2i0.21927
Kim WS et al (2009) Antiwrinkle effect of adipose-derived stem cell: activation of dermal fibroblast by secretory factors. J Dermatol Sci 53:96–102
Kooijmans SA et al (2012) Exosome mimetics: a novel class of drug delivery systems. Int J Nanomedicine 7:1525–1541
Lamichhane TN et al (2015) Emerging roles for extracellular vesicles in tissue engineering and regenerative medicine. Tissue Eng B Rev 21:45–54
Lane RE et al (2015) Analysis of exosome purification methods using a model liposome system and tunable-resistive pulse sensing. Sci Rep 5:7639
Lee C et al (2012) Exosomes mediate the cytoprotective action of mesenchymal stromal cells on hypoxia induced pulmonary hypertension. Circulation 126:2601–2611
Lee HJ et al (2014) Efficacy of microneedling plus human stem cell conditioned medium for skin rejuvenation: a randomized, controlled, blinded split face study. Ann Dermatol 26:584–591
Lener T et al (2015 Dec 31) Applying extracellular vesicles based therapeutics in clinical trials - an ISEV position paper. J Extracell Vesicles 4:30087
Li M et al (2015) Mesenchymal stem cell-conditioned medium improves the proliferation and migration of keratinocytes in a diabetes-like microenvironment. Int J Low Extrem Wounds 14:73
Liegeouis S et al (2006) The V0-ATPase mediates apical secretion of exosomes containing Hedgehog-related proteins in *Caenorhabditis elegans*. J Cell Biol 173:949–961
Lim CP et al (2006) Stat3 contributes to keloid pathogenesis via promoting collagen production, cell proliferation, and migration. Oncogene 25:5416–5425

Liu JJ et al (2013) Preparation and characterization of cosmeceutical liposomes loaded with avobenzone and arbutin. J Cosmet Sci 64:9–17
Lötvall J et al (2014 Dec 22) Minimal experimental requirements for definition of extracellular vesicles and their functions: a position statement from the International Society for Extracellular Vesicles. J Extracell Vesicles 3:26913
Machova Urdzikova L et al (2014) Human multipotent mesenchymal stem cells improve healing after collagenase tendon injury in the rat. Biomed Eng Online 13:42
Madsen JT, Andersen KE (2010) Microvesicle formulations used in topical drugs and cosmetics affect product efficiency, performance and allergenicity. Dermatitis 21:243–247
Maguire G (2013) Stem cell therapy without the cells. Commun Integr Biol 6:e26631
Marcus ME, Leonard JN (2013) FedExosomes: engineering therapeutic biological nanoparticles that truly deliver. Pharmaceuticals 6:659–680
Melo SA et al (2014) Cancer exosomes perform cell-independent microRNA biogenesis and promote tumorigenesis. Cancer Cell 26:707–721
Morse MA et al (2005) A Phase I study of dexosome immunotherapy in patients with advanced non-small cell lung cancer. Clin Cancer Res 11:3017–3024
Mu X et al (2013) Regeneration of soft tissues is promoted by MMP1 treatment after digit amputation in mice. PLoS One 8:e59105
Pan BT et al (1985) Electron microscopic evidence for externalization of the transferrin receptor in vesicular form in sheep reticulocytes. J Cell Biol 101:942–948
Quesenberry PJ et al (2014) Cellular phenotype and extracellular vesicles: basic and clinical considerations. Stem Cells Dev 23:1429–1436
Rahimpour Y, Hamishehkar H (2012) Liposomes in cosmeceutics. Expert Opin Drug Deliv 9:443–455
Ranghino A et al (2015) Extracellular vesicles in the urine: markers and mediators of tissue damage and regeneration. Clin Kidney J 8:23–30
Rashed H et al (2017) Exosomes: from garbage bins to promising therapeutic targets. Int J Mol Sci 18(3). pii: E538
Ratajczak MZ et al (2012) Pivotal role of paracrine effects in stem cell therapies in regenerative medicine: can we translate stem cell-secreted paracrine factors and microvesicles into better therapeutic strategies? Leukemia 26:1166–1173
Roy S, Gatien S (2008) Regeneration in axolotls: a model to aim for! Exp Gerontol 43:968–973
Sahoo S et al (2011) Exosomes from human CD34+ stem cells mediate their proangiogenic paracrine activity. Circ Res 109:724–728
Sen CK et al (2009) Human skin wounds: a major and snowballing threat to public health and the economy. Wound Repair Regen 17:763
Seo KY et al (2013) Skin rejuvenation by microneedle fractional radiofrequency and a human stem cell conditioned medium in Asian skin: a randomized controlled investigator blinded split face study. J Cosmet Laser Ther 15:25–33
Shim JH et al (2013) Human dermal stem/progenitor cell-derived conditioned medium ameliorates ultraviolet a induced damage of normal human dermal fibroblasts. PLoS One 8:e67604
Tanaka EM, Reddien PW (2011) The cellular basis for animal regeneration. Dev Cell 21:172–185
Tanaka Y et al (2005) FGF-induced vesicular release of Sonic hedgehog and retinoic acid in leftward nodal flow is critical for left-right determination. Nature 435:172–177
Tetta C et al (2012) The role of micro-vesicles derived from mesenchymal stem cells in tissue regeneration; a dream for tendon repair? Muscles Ligaments Tendons J 2:212–221
Valadi H et al (2007) Exosome mediated transfer of mRNAs and microRNAs is a novel mechanism of genetic exchange between cells. Nat Cell Biol 9:654–659
van Koppen A et al (2012) Human embryonic mesenchymal stem cell-derived conditioned medium rescues kidney function in rats with established chronic kidney disease. PLoS One 7:e38746
Vyas N et al (2014) Vertebrate Hedgehog is secreted on two types of extracellular vesicles with different signaling properties. Sci Rep 4:7357
Wang CY et al (2012) Mesenchymal stem cell-conditioned medium facilitates angiogenesis and fracture healing in diabetic rats. J Tissue Eng Regen Med 6:559

Wang J et al (2017) Exosomes: a novel strategy for treatment and prevention of diseases. Front Pharmacol 8:300

Wessels NK (1977) Tissue interactions and development. Benjamin Cummings, Menlo Park, CA

Xin H et al (2013) Systematic administration of exosomes released from mesenchymal stromal cells promote functional recovery and neurovascular plasticity after stroke in rats. J Cereb Blood Flow Metab 33:1711–1715

Yu B et al (2014) Exosomes derived from mesenchymal stem cells. Int J Mol Sci 15:4142–4157

Zhang HG, Grizzle WE (2014) Exosomes: a novel pathway of local and distant intercellular communication that facilitates the growth and metastasis of neoplastic lesions. Am J Pathol 184:28–41

Zhang B et al (2014) HucMSC-exosome mediated -Wnt4 signaling is required for cutaneous wound healing. Stem Cells. https://doi.org/10.1002/stem.1771

Zhang J et al (2015a) Exosomes released from human induced pluripotent stem cells derived MSCs facilitate cutaneous wound healing by promoting collagen synthesis and angiogenesis. J Transl Med 13:49

Zhang Y et al (2015b) Effect of exosomes derived from multipluripotent mesenchymal stromal cells on functional recovery and neurovascular plasticity in rats after traumatic brain injury. J Neurosurg 122:856–867

Chapter 5
Proinflammatory Cytokines Significantly Stimulate Extracellular Vesicle Production by Adipose-Derived and Umbilical Cord-Derived Mesenchymal Stem Cells

Phuc Van Pham, Ngoc Bich Vu, Khanh Hong-Thien Bui, and Liem Hieu Pham

Introduction

Extracellular vesicles (EVs) are nanosized particles produced from live cells during their life span. Depending on their size, they are generally divided into two main groups: exosomes (40–150 nm in diameter) and microvesicles (50–2000 nm in diameter). Some studies have suggested that there are four kinds of EVs: exosomes, microvesicles, apoptotic bodies, and oncosomes. These EVs play an important role in cell–cell communication. Indeed, EVs contain a broad range of biological molecules, from DNA to RNA (Eirin et al. 2014; Kumar et al. 2015; Vallabhaneni et al. 2015) to proteins (Baglio et al. 2012; Biancone et al. 2012; Rani et al. 2015).

Since EVs transport DNA, RNA, and proteins, they presumably have the capability to regulate target cells—from transcription and translation—and thus have been evaluated for the treatment of various diseases. Such diseases include kidney disease, heart disease, liver disease, and brain injury. In kidney disease, EVs have been mainly used to treat acute kidney injury (AKI). EVs derived from mesenchymal stem cells (i.e., MSC-EVs) have been shown to induce significant

P. Van Pham (✉) · N. B. Vu
Laboratory of Stem Cell Research and Application, VNUHCM University of Science, Ho Chi Minh City, Vietnam

Stem Cell Institute, VNUHCM University of Science, Ho Chi Minh City, Vietnam
e-mail: pvphuc@hcmuns.edu.vn; phucpham@sci.edu.vn

K. H.-T. Bui
University Medical Center, University of Medicine and Pharmacy, Ho Chi Minh City, Vietnam

L. H. Pham
Pham Ngoc Thach University of Medicine, Ho Chi Minh City, Vietnam

P. V. Pham (ed.), *Stem Cell Drugs - A New Generation of Biopharmaceuticals*, Stem Cells in Clinical Applications, https://doi.org/10.1007/978-3-319-99328-7_5

improvement of kidney function (Bruno et al. 2012; Hu et al. 2016). Some mechanisms of action of MSC-EVs include inhibiting oxidative stress, apoptosis, and fibrosis (Lin et al. 2016; Zhou et al. 2013; Zou et al. 2014, 2016), stimulating angiogenesis (Lin et al. 2016; Ranghino et al. 2017), and mediating an anti-inflammatory condition (Koch et al. 2015; Lin et al. 2016). In heart disease, MSC-EVs have also shown some promising benefits in acute myocardial infarction (AMI) models. MSC-EV infusion has helped to reduce infarction size and heart function in animal models of AMI (Ma et al. 2016; Shao et al. 2017; Teng et al. 2015). Similar to kidney disease treatment, EVs were shown to also reduce fibrosis and apoptosis, and stimulate angiogenesis (Feng et al. 2014; Yu et al. 2016; Zhao et al. 2015). Moreover, EVs have been evaluated for the treatment of liver injuries/diseases, such as hepatic failure and hepatic ischemia, and were shown to induce improvement of liver function (Chen et al. 2017; Haga et al. 2017; Nong et al. 2016; Tan et al. 2014).

Recently, some studies have evaluated MSC-EV therapy for brain injuries, such as ischemic stroke and traumatic brain injury (Doeppner et al. 2015; Kim et al. 2016; Zhang et al. 2016). MSC-EVs have also been used to treat hippocampal synaptic impairment after transient global ischemia (Deng et al. 2017), fetal brain injury after hypoxia ischemia (Ophelders et al. 2016), post-stroke conditions (Doeppner et al. 2015), and cerebral apoplexy (Hu et al. 2016). Deng et al. (2017) showed that MSC-EVs significantly inhibited ischemia-induced pathogenic expression of COX-2 in the hippocampus and ameliorated effects on synaptic functions. Ophelders et al. (2016) used MSC-EVs from bone marrow to treat hypoxic ischemic injury of the preterm brain, and showed that both function and structural injury of the fetal brain improved (Ophelders et al. 2016). MSC-EVs also induce neuroprotection and neural regeneration (Doeppner et al. 2015; Hu et al. 2016).

However, to date there have only been two clinical trials using MSC-EVs. The first trial using MSC-EVs was for graft-versus-host disease (GVHD). The initial results showed that MSC-EV therapy was helpful for controlling GVHD (and its symptons) as well as reducing the dose of steroids (Kordelas et al. 2014). The second study examined the use of MSC-EVs to treat chronic kidney disease. The results also showed that MSC-EV transplantation significantly improved kidney function compared to controls (Nassar et al. 2016).

Nonetheless, the main limitation of MSC-EVs in clinical applications is the low quantity of EVs obtained from MSC cultures. A recent study by Lo Sicco et al. (2017) showed that ADSCs can produce about 20 EVs/μL in the conditioned medium under normoxia and 35 EVs/μL under hypoxia (Lo Sicco et al. 2017). Therefore, boosting MSCs to produce more EVs has become an important focus in the clinic. This study aims to enhance EV production by ADSCs and UC-MSCs via the use of proinflammatory cytokines (TNF-alpha and IFN-gamma). Results from this study will significantly contribute to clinical applications of MSC-EVs.

Materials and Methods

Culture and Expansion of ADSCs

Adipose tissues were collected from the hospital from aesthetic surgery with consent from the donors. The tissues were kept in saline and transferred to the laboratory. The use and manipulation of adipose tissues were approved by the institutional ethics committee. In the laboratory, the adipose tissues were washed twice with PBS to remove blood cells and then extracted to obtain stromal vascular fractions (SVF) with a commercial kit (Cell Extraction Kit, Regenmed Ltd., Ho Chi Minh City, Vietnam). Briefly, clean adipose tissues were incubated with collagenase enzyme (SuperDigest Enzyme, Regenmed Ltd., Ho Chi Minh City, Vietnam) for 15 min with extractor. Finally, the digested tissues were centrifuged at 3500 rpm for 15 min to collect SVFs at the bottom of 50-mL Falcon tubes. SVFs were resuspended into MSCCult Medium (Regenmed Ltd., Ho Chi Minh City, Vietnam), which consisted of 90% DMEM/F12, 10% FBS, and 1% antibiotic–antimycotic solution (all purchased from Thermo Fisher Scientific Inc., Waltham, MA). SVFs were cultured under standard conditions (37 °C and 5% CO_2). After the ADSC culture reached about 70% confluence, the cells were subcultured to the third passage for use in experiments.

Culture and Expansion of UC-MSCs

The culture and expansion of UC-MSCs were carried out according to previously published protocols (Van Pham et al. 2016b). Briefly, umbilical cords were collected from the hospital (Van Hanh Hospital, Ho Chi Minh City, VN) with signed consent forms from donors. The samples were transferred to the laboratory within 2 h. There, they were washed twice with PBS and cut into small size pieces (1–2 mm^2). The fragments were seeded onto the surface of flasks (T-75 flask; SPL, Korea) using a drop of MSCCult medium and left for 5 days. After 5 days, the MSC candidates migrated from the fragments. Fresh medium (up to 12 mL) was added to the flasks and cells were monitored. After the UC-MSC culture reached about 70% confluence, the cells were subcultured to the third passage for use in experiments.

In Vitro Differentiation

MSC candidates were induced to differentiate into adipocytes and osteoblasts according to previously published protocols (Van Pham et al. 2014, 2016a). Briefly, they were cultured in the 6-well plates (SPL, Korea) until they reached 50% confluence. Differentiation medium (Thermo Fisher Inc., Waltham, MA) was added to the plates. The medium was replenished every 3 days for a duration of 21 days (from the first day of induction). The cells were stained with Oil Red (to confirm for adipocytes) and with Alizarin Red (to confirm for osteoblasts).

Marker Analysis of MSCs

The markers of MSCs were evaluated based on flow cytometry. Firstly, the MSC candidates were detached from the flasks and resuspended in FACSFlow sheath fluid at 10^6 cells/100 μL solution. They were stained individually with anti-CD14-FITC, anti-CD45-APC, anti-CD44-PE, anti-CD73-PerCP, and anti-CD90-FITC antibodies in FACS tubes for 20 min at room temperature (an isotype antibody was used for the control tube). The tubes were added up to 300 μL with FACSFlow sheath fluid before they were run on the FACSCalibur machine (BD Biosciences, San Jose, CA). Analysis of marker expression was performed using CellQuest Pro software using 10^4 cells/events per marker.

Culture and Isolation of EVs from MSCs

After confirmation of MSC markers, ADSCs and UC-MSCs were cultured under standard conditions until they reached about 70% confluence (described above). At this point, ADSCs and UC-MSCs were collected from culture using standard trypsin/EDTA-0.04% solution. Cells were washed twice with PBS, and 12 mL of fresh culture medium (DMEM/F12 supplemented with exosome-depleted FBS and 1% antibiotic–antimycotic solution) was added as described above. TNF-alpha and IFN-gamma (Santa Cruz Biotechnology, Mississauga, ON, Canada) were added to some cell cultures in fresh medium (starting point, 0 h) at various concentrations: 5, 10, and 20 ng/mL. After 24, 48, and 96 h of incubation, the conditioned media were collected to prepare EVs. The conditioned medium was centrifuged at 2500 × *g* for 20 min at 4 °C to remove cell debris and large vesicles; EVs were isolated from the supernatant using Total Exosome Isolation Reagent Kit (Thermo Fisher Scientific Inc.).

EV Characterization and Quantification

EV characterization and quantification was performed using a published protocol (Di Trapani et al. 2016). Briefly, the flow cytometer was calibrated using different fluorescent beads (Thermo Fisher Scientific Inc.) of various sizes (0.1, 0.2, 0.5, and 1.0 μm). The beads were mixed with EVs to generate an analytic gate for the subsequent experiments. EV quantification was obtained by Trucount tubes (BD Biosciences, San Jose, CA) to obtain the absolute numbers. The tubes were used according to the manufacturer's recommendations and the absolute count was calculated using the following formula: (number of events in the EV-containing gate/number of events in the bead-containing gate) × (number of beads per test/volume).

For phenotypic analysis of EVs, the EVs were adsorbed to 3.9 μm latex beads (Thermo Fisher Scientific). EVs were mixed with latex beads for 15 min at room temperature. The bead-bound EVs were collected by centrifuging for 3 min at 4000 rpm. After washing the pellets in PBS/0.5% BSA, they were resuspended in 0.5 mL of PBS/0.5% BSA. Finally, 10 μL of bead-bound EVs was stained with specific antibodies (anti-CD44, -CD73, -CD90, -CD63, and -CD81) for 30 min at room temperature. Data were collected and analyzed in the FACSCalibur Machine using 100,000 events.

Transmission Electron Microscopy

EVs were fixed in 4% paraformaldehyde solution. The images were captured using a transmission electron microscopy using a digital Morada G2 TEM camera (Olympus Imaging Systems, Japan).

Statistical Analysis

Data are presented as the mean ± standard deviation (SD). Prism Version 6.00 for Mac (GraphPad Software, La Jolla, CA) was used for statistical analysis. Unpaired t-tests were performed; p-values <0.05 were considered statistically significant.

Results

Isolation and Characterization of ADSCs

After 24 h of incubation (as recommended by a published protocol for isolating ADSCs), some spindle-shaped cells appeared which were clearly visible under a microscope. These unique cells grow and reached 70% confluence on the surface of flasks in about 7 days. These cells were subcultured and they became even more homogenous by the third passage and were then collected for use in experiments.

To confirm their MSC phenotype, ADSCs were assessed for the following surface markers: CD44, CD73, CD90, CD14, and CD45. The results showed that ADSCs were strongly positive for CD44 (99.21 ± 1.21%), CD73 (95 ± 5.4%), CD90 (100%), while negative for CD14 (2.1 ± 1.4%) and CD45 (4.1 ± 1.4%). During induction of adipogenesis, cells became differentiated into adipocytes. Indeed, these adipocytes stained positive with Oil Red staining assay. During induction of osteoblast differentiation, cells could be differentiated into osteoblasts and, indeed, stained positive with Alizarin Red staining assay.

Isolation and Characterization of UC-MSCs

UC-MSCs were isolated from tissues and expanded in tissue culture flasks. After 4 days of incubation, spindle-shape cells migrated from the tissue. After 14 days of culture, the cell population reached 70% confluence. Cells became homogenous after the third passage. UC-MSCs were also confirmed to be MSCs from their expression of CD44 (98.11 ± 3.23%), CD73 (96.81 ± 3.11%), and CD90 (99.91 ± 0.11%); they were negative for CD14 (2.01 ± 0.81%) and CD45 (3.51 ± 2.19%). These cells were succesfully induced to differentiate into adipocytes and osteoblasts. The results showed that the obtained UC-MSC candidates successfully differentiated into both adipocytes and osteoblasts.

UC-MSCs and ADSCs Can Produce EVs Expressing Some Standard Markers of MSCs

EVs from both UC-MSCs and ADSCs were evaluated for expression of some MSC surface markers (CD44, CD73, and CD90) and some EV markers (CD81 and CD63). The EVs were also visualized and the images captured under TEM to determine their diameter. The results showed that almost all EVs expressed some common markers of MSC-EVs, such as CD44 (99.21 ± 1.21%), CD73 (99.21 ± 1.21%), and CD90 (99.21 ± 1.21%). From TEM capture, the vesicles were estimated to be 30–200 nm in diameter.

TNF-Alpha Stimulates UC-MSCs and ADSCs to Produce EVs Depending on Its Concentration and Time of Incubation

The results in Fig. 5.1 showed that increasing concentrations of TNF-alpha boosted both UC-MSC and ADSCs to produce more EVs. For ADSCs, when the concentration of TNF-alpha increased from 5 to 10 and 20 ng/mL, the EVs counts significantly increased—from 200 ± 30 to 353.33 ± 32.15 and 450 ± 50, respectively (after 24 h of incubation), from 350 ± 50 to 590 ± 36.06 and 740 ± 65.57, respectively (after 48 h of incubation), and from 210 ± 52.92 to 400 ± 100 and 466.67 ± 76.38, respectively (after 96 h of incubation) ($n = 3$; $p < 0.05$ for all sets). For UC-MSCs, similarly when concentrations of TNF-alpha increased from 5 to 10 and 20 ng/mL, the EVs counts significantly increased—from 45 ± 5 to 88.33 ± 7.64 and 100 ± 10,

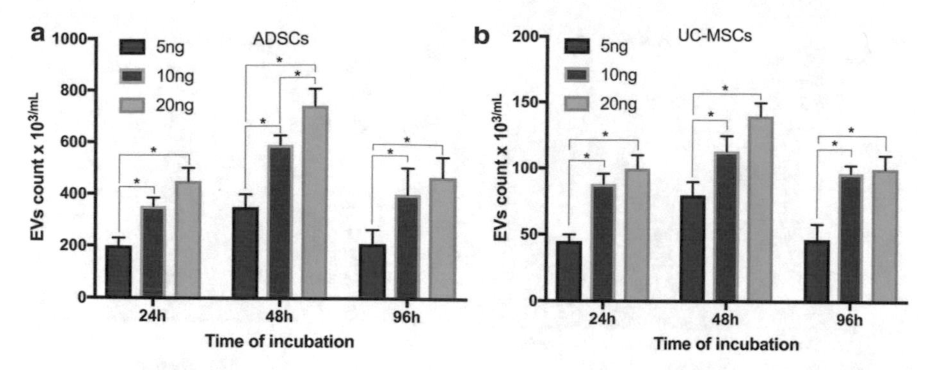

Fig. 5.1 TNF-alpha stimulates UC-MSCs and ADSCs to produce EVs. EV production by ADSCs (**a**) and UC-MSCs (**b**) gradually increased as the concentration of TNF-alpha increased. However, the EV count for ADSCs (**a**) and UC-MSCs (**b**) was maximal at 48 h of incubation with TNF-alpha (compared to 24 or 96 h)

respectively (after 24 h of incubation), from 80 ± 10 to 113.33 ± 11.55 and 140 ± 10, respectively (after 48 h of incubation), and from 46.67 ± 11.55 to 96.67 ± 5.77 and 100 ± 10, respectively (after 96 h of incubation) ($n = 3$; $p < 0.05$ for all sets).

Moreover, the EV counts produced by the UC-MSCs and ADSCs were dependent on the duration of incubation with TNF-alpha. The greatest EV count for ADSCs and UC-MSCs occurred after 48 h of incubation with TNF-alpha (as compared to 24 or 96 h) (Fig. 5.1). Indeed, at 5 ng/mL of TNF-alpha, EVs counts for ADSCs were 200 ± 30, 350 ± 50, and 210 ± 52.92, respectively (for 24, 48, and 96 h of incubation). Similarly, at 10 ng/mL of TNF-alpha, EVs counts for ADSCs were 353.33 ± 32.15, 590 ± 36.06, and 400 ± 100, respectively (for 24, 48, and 96 h of incubation). Thus, the EV count was highest after 48 h of incubation and gradually declined after 96 h of incubation. At 20 ng/mL of TNF-alpha, EVs counts for ADSCs were 450 ± 50, 740 ± 65.57, and 466.67 ± 76.38, respectively (for 24, 48, and 96 h of incubation); thus, EV count was optimal at 48 h. The time-dependent trend for UC-MSCs was also similar to that of ADSCs, with 48 h being the optimal incubation time for generating EVs (Fig. 5.1).

IFN-Gamma Stimulates UC-MSCs and ADSCs to Produce EVs Depending on Its Concentration and Time of Incubation

Similarly to TNF-alpha, IFN-gamma also stimulated both ADSCs and UC-MSCs to produce more EVs in a dose-dependent and time dependent manner. In general, as the concentration of IFN-gamma increased, the EV counts for both ADSCs and UC-MSCs increased. For ADSCs, when the concentration of IFN-gamma increased from 5 to 10 and 20 ng/mL, the EVs counts significantly increased—from 300 ± 20 to 486.67 ± 32.15 and 900 ± 100, respectively (after 24 h of incubation) ($n = 3$; $p < 0.05$), from 396.67 ± 25.166 to 863.33 ± 56.86 and 1063.33 ± 118.46,

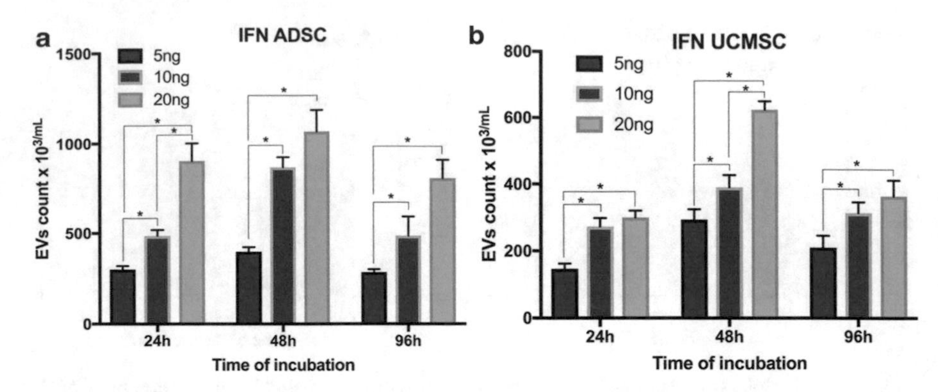

Fig. 5.2 IFN-gamma stimulates UC-MSCs and ADSCs to produce EVs. EV production by ADSCs (**a**) and UC-MSCs (**b**) gradually increased as the concentration of IFN-gamma increased. However, the EV count for ADSCs (**a**) and UC-MSCs (**b**) was maximal at 48 h of incubation with IFN-gamma (compared to 24 or 96 h)

respectively (after 48 h of incubation) ($n = 3$; $p < 0.05$), and from 283.33 ± 15.28 to 483.33 ± 104.08 and 800 ± 100, respectively ($n = 3$; $p < 0.05$). Similarly, for UC-MSCs, the EV counts also increased depending on the concentration of IFN-gamma—from 5 ng/mL (146.67 ± 15.28, 293.33 ± 30.55, and 210 ± 36.056, respectively, after 24, 48, and 96 h) to 10 ng/mL (273.33 ± 25.17, 390 ± 36.06, and 313.33 ± 32.15, respectively, after 24, 48, and 96 h) and to 20 ng/mL (300 ± 20, 623.33 ± 25.17, and 363.33 ± 47.26, respectively, after 24, 48, and 96 h).

Like TNF-alpha, as IFN-gamma incubation time increased from 24 to 48 h there was an increase of EV count; however, when incubation time was increased to 96 h the EV counts were reduced. For ADSCs, EV counts at 24 h (300 ± 20, 486.67 ± 32.15, and 900 ± 100, respectively, for 5, 10, and 20 ng/mL of IFN-gamma) increased after 48 h (396.67±25.166, 863.33 ± 56.86, and 1063.33 ± 118.46, respectively, for 5, 10, and 20 ng/mL of IFN-gamma), but decreased after 96 h (283.33 ± 15.28, 483.33 ± 104.08 and 800 ± 100, respectively, for 5, 10, and 20 ng/mL of IFN-gamma). For UC-MSCs, EV counts at 24 h (146.67 ± 15.28, 273.33 ± 25.17, and 300 ± 20, respectively, for 5, 10, and 20 ng/mL of IFN-gamma) increased after 48 h (293.33 ± 30.55, 390 ± 36.06, and 623.33 ± 25.17, respectively, for 5, 10, and 20 ng/mL of IFN-gamma), but decreased after 96 h (210 ± 36.056, 313.33 ± 32.15, and 363.33 ± 47.26, respectively, for 5, 10, and 20 ng/mL of IFN-gamma) (Fig. 5.2).

IFN-Gamma Stimulates a Stronger Induction of EV Production by UC-MSCs and ADSCs than TNF-Alpha

We compared the effects of TNF-alpha and IFN-gamma on EV production. The EV counts produced by ADSCs and UC-MSCs for the same time of incubation and same concentration (of TNF-alpha and IFN-gamma) were recorded (Fig. 5.3).

As presented in Fig. 5.3, when comparing the same concentration of TNF-alpha and IFN-gamma, and same time of incubation, IFN-gamma was more robust at stimulating ADSCs and UC-MSCs to produce EVs. However, the effects of IFN-gamma were stronger (and more evident) on UC-MSCs than ADSCs.

For ADSCs, the comparison of EV counts induced by IFN-gamma versus TNF-alpha were as follows: 300 ± 20 vs 200 ± 30, 486.67 ± 32.15 vs 353.33 ± 32.15, and 900 ± 100 vs. 450 ± 50 (for 5, 10, and 20 ng/mL, respectively, IFN-gamma vs. TNF-alpha, after 24 h); 396.67 ± 25.17 vs 350 ± 50, 863.33 ± 56.86 vs 590 ± 36.06, and 1063.33 ± 118.46 vs 740 ± 65.57 (for 5, 10, and 20 ng/mL, respectively, IFN-gamma vs. TNF-alpha, after 48 h); and 283.33 ± 15.28 vs. 210 ± 52.92, 483.33 ± 104.08 vs. 400 ± 100, and 800 ± 100 vs 466.67 ± 76.38 (for 5, 10, and 20 ng/mL, respectively, IFN-gamma vs. TNF-alpha, after 96 h) (Fig. 5.3).

For UC-MSCs, the comparison of EV counts induced by IFN-gamma versus TNF-alpha was as follows: 146.67 ± 15.28 vs. 45 ± 5, 273.33 ± 25.17 vs. 88.33 ± 7.64, and 300 ± 20 vs. 100 ± 10 (for 5, 10, and 20 ng/mL, respectively, IFN-gamma vs. TNF-alpha, after 24 h); 293.33 ± 30.55 vs. 80 ± 10, 390 ± 36.056 vs. 113.33 ± 11.55, and 623.33 ± 25.17 vs. 140 ± 10 (for 5, 10, and 20 ng/mL, respectively, IFN-gamma vs. TNF-alpha, after 48 h); and 210 ± 36.06 vs. 46.67 ± 11.55, 313.33 ± 32.15 vs. 96.67 ± 5.77, and 363.33 ± 47.26 vs. 100 ± 10 (for 5, 10, and 20 ng/mL, respectively, IFN-gamma vs. TNF-alpha, after 96 h) (Fig. 5.3).

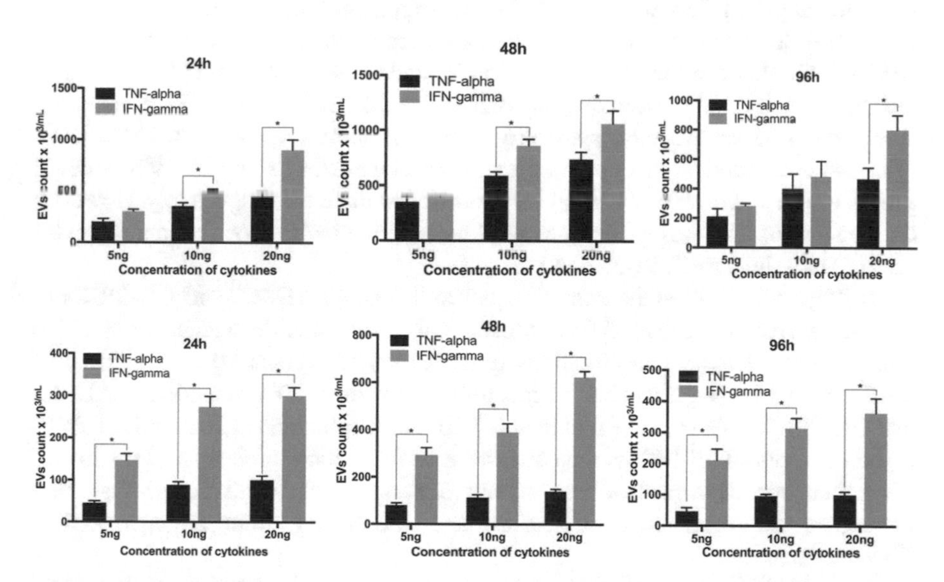

Fig. 5.3 IFN-gamma stimulates a stronger induction of EV production by UC-MSCs and ADSCs than TNF-alpha. At the same cytokine concentration, EV production induced by IFN-gamma treatment was nearly twofold greater than that induced by TNF-alpha

Discussion

MSC

EVs have become a new platform for cell-free therapy for some diseasesespecially degenerative diseases. These particles have been evaluated for the treatment of various diseases including kidney, brain, heart and liver diseases. More importantly-clinical applications of these particles in GVHD and chronic kidney disease treatment have indicated that MSC-EVs are promising candidates to replace MSC transplantation.

In this study we investigated the effects of TNF-alpha and IFN-gamma on in vitro EV production by ADSCs and UC-MSCs. In this initial study we show that both TNF-alpha and IFN-gamma can significant enhance EV production by ADSCs and UC-MSCs. Firstly, we successfully isolated MSCs from adipose tissue and umbilical cord. These cells satisfied the criteria of MSCs posed by Dominici et al. (2006). Indeed, they could adhere onto plastic flasks and exhibited fibroblast-like shape; they also expressed some common markers of MSCs (e.g., CD44, CD73, and CD90) as well as being negative for blood cell markers (e.g., CD14 and CD45). They also could be induced to differentiate into various kinds of mesoderm-induced adipocytes and osteoblasts. The phenotype of ADSCs and UC-MSCs were similar to previous studies (Van Pham et al. 2014, 2016a, b).

Secondly, these MSCs were capable of producing EVs in culture medium with exosome-depleted FBS supplement. EVs were regarded in this setting as placebo. EVs generated from MSCs (MSC-EVs) exhibited all the specific characteristics of MSCs, including expression of CD44, CD73, and CD90, as confirmed to be important by previous studies (Ramos et al. 2016). The diameters of MSC-EVs were confirmed by TEM capture. Flow cytometry also confirmed these particles as MSC-EVs (Tru-count determined that under normal conditions ADSCs and UC-MSCs could produce 84 ± 14 and 30 ± 9 EVs/μL of conditioned medium, respectively. However, the quantity of EVs was further increased by treatment with proinflammatory cytokines (TNF-alpha and IFN-gamma).

Thirdly, both TNF-alpha and IFN-gamma triggered ADSCs and UC-MSCs to produce a greater number of EVs. Indeed, this effect was dependent on cytokine concentration. Higher concentrations (e.g., 20 ng/mL versus 10 or 5 ng/mL) of TNF-alpha and IFN-gamma led to a greater stimulation of EV production by ADSCs and UC-MSCs. However, the maximal EV count was obtained at 20 ng/mL of TNF-alpha or 20 ng/mL of IFN-gamma after 48 h of incubation; after 96 h of treatment, EV production decreased. Moreover, our results demonstrated that IFN-gamma stimulated EV production of ADSCs and UC-MSCs more effectively than TNF-alpha.

From the literature, EV production can be enhanced by hypoxia treatment of MSCs (Lo Sicco et al. 2017). From hypoxia treatment of ADSCs, Lo Sicco et al. (2017) showed that EV quantity doubled. However, there have been no reports in the literature comparing TNF-alpha- or IFN-gamma-mediated effects on EV production.

Some published studies have shown that these inflammatory factors stimulate secretion by MSCs. By treating with TNF-alpha and IFN-gamma, MSCs can increase their production of VEGF, HGF, IDO, TGF-beta, PGE2, BMP2, Factor H, Gal-9, and TSG-6 (Madrigal et al. 2014). Xing et al. (2014) showed that proinflammatory factors, including IL-1β, IL-6, and TNF-α, could promote bone marrow MSCs to increase chemokine secretion (Xing et al. 2014). Besides the secretory processes, TNF-alpha have also been shown to be stimulator of MSC activities, and TNF-alpha pretreatment of MSCs can induce superior angiogenic activity in vitro compared to untreated MSCs (Kwon et al. 2013). Moreover, TNF-alpha pretreated MSCs have shown enhanced proliferation, mobilization, and osteogenic differentiation (Lu et al. 2013). Although the mechanisms by which TNF-alpha and IFN-gamma mediate their effects on EV production by ADSCs and UC-MSCs were not demonstrated in this study, based on published studies we postulate that their beneficial effects are related (directly or indirectly) to processes involved in production.

Conclusion

Extracellular vesicles (EVs) are particles, ranging from nanometers to micrometers in size, which are produced by live cells including MSCs. These particles contain various biological molecules, such as siRNA, RNA, DNA, and proteins, which can regulate target cells. As such, EVs have been used in treating various degenerative diseases and have shown promising results. The study herein demonstrates a new method to enhance EV production from ADSCs and UC-MSCs by treatment/culture with TNF-alpha or IFN-gamma.

The results showed that at 20 ng/mL of TNF-alpha (as well as IFN-gamma) and after 24 h of incubation, both ADSCs and UC-MSCs can produce the maximal quantity of EVs. This study also shows that IFN-gamma can boost MSCs to produce EVs more effectively than TNF-alpha. These results suggest that pretreatment of MSCs with IFF gamma may be an effective strategy to optimize EV use in clinical applications.

Acknowledgment This research was funded and supported by Vietnam National University Ho Chi Minh City via project TX2017-18-02, by Ministry of Science and Technology, Vietnam under grant number DM.10.DA/15; by Fostering Innovation through Research, Science and Technology, Vietnam via project 15/FIRST/2a/SCI.

References

Baglio SR, Pegtel DM, Baldini N (2012) Mesenchymal stem cell secreted vesicles provide novel opportunities in (stem) cell-free therapy. Front Physiol 3:359

Biancone L, Bruno S, Deregibus MC, Tetta C, Camussi G (2012) Therapeutic potential of mesenchymal stem cell-derived microvesicles. Nephrol Dial Transplant 27:3037–3042

Bruno S, Grange C, Collino F, Deregibus MC, Cantaluppi V, Biancone L, Tetta C, Camussi G (2012) Microvesicles derived from mesenchymal stem cells enhance survival in a lethal model of acute kidney injury. PLoS One 7:e33115

Chen L, Xiang B, Wang X, Xiang C (2017) Exosomes derived from human menstrual blood-derived stem cells alleviate fulminant hepatic failure. Stem Cell Res Ther 8:9

Deng M, Xiao H, Zhang H, Peng H, Yuan H, Xu Y, Zhang G, Hu Z (2017) Mesenchymal stem cell-derived extracellular vesicles ameliorates hippocampal synaptic impairment after transient global ischemia. Front Cell Neurosci 11:205

Di Trapani M, Bassi G, Midolo M, Gatti A, Kamga PT, Cassaro A, Carusone R, Adamo A, Krampera M (2016) Differential and transferable modulatory effects of mesenchymal stromal cell-derived extracellular vesicles on T, B and NK cell functions. Sci Rep 6:24120

Doeppner TR, Herz J, Gorgens A, Schlechter J, Ludwig AK, Radtke S, de Miroschedji K, Horn PA, Giebel B, Hermann DM (2015) Extracellular vesicles improve post-stroke neuroregeneration and prevent postischemic immunosuppression. Stem Cells Transl Med 4:1131–1143

Dominici M, Le Blanc K, Mueller I, Slaper-Cortenbach I, Marini F, Krause D, Deans R, Keating A, Prockop D, Horwitz E (2006) Minimal criteria for defining mulipotent mesenchymal stromal cells. The International Society for Cellular Therapy position statement. Cytotherapy 8:315–317

Eirin A, Riester SM, Zhu XY, Tang H, Evans JM, O'Brien D, van Wijnen AJ, Lerman LO (2014) MicroRNA and mRNA cargo of extracellular vesicles from porcine adipose tissue-derived mesenchymal stem cells. Gene 551:55–64

Feng Y, Huang W, Wani M, Yu X, Ashraf M (2014) Ischemic preconditioning potentiates the protective effect of stem cells through secretion of exosomes by targeting Mecp2 via miR-22. PLoS One 9:e88685

Haga H, Yan IK, Takahashi K, Matsuda A, Patel T (2017) Extracellular vesicles from bone marrow-derived mesenchymal stem cells improve survival from lethal hepatic failure in mice. Stem Cells Transl Med 6:1262–1272

Hu B, Chen S, Zou M, He Z, Shao S, Liu B (2016) Effect of extracellular vesicles on neural functional recovery and immunologic suppression after rat cerebral apoplexy. Cell Physiol Biochem 40:155–162

Kim DK, Nishida H, An SY, Shetty AK, Bartosh TJ, Prockop DJ (2016) Chromatographically isolated CD63+CD81+ extracellular vesicles from mesenchymal stromal cells rescue cognitive impairments after TBI. Proc Natl Acad Sci U S A 113:170–175

Koch M, Lemke A, Lange C (2015) Extracellular vesicles from MSC modulate the immune response to renal allografts in a MHC disparate rat model. Stem Cells Int 2015:486141

Kordelas L, Rebmann V, Ludwig AK, Radtke S, Ruesing J, Doeppner TR, Epple M, Horn PA, Beelen DW, Giebel B (2014) MSC-derived exosomes: a novel tool to treat therapy-refractory graft-versus-host disease. Leukemia 28:970–973

Kumar L, Verma S, Vaidya B, Gupta V (2015) Exosomes: natural carriers for siRNA delivery. Curr Pharm Des 21:4556–4565

Kwon YW, Heo SC, Jeong GO, Yoon JW, Mo WM, Lee MJ, Jang IH, Kwon SM, Lee JS, Kim JH (2013) Tumor necrosis factor-alpha-activated mesenchymal stem cells promote endothelial progenitor cell homing and angiogenesis. Biochim Biophys Acta 1832:2136–2144

Lin KC, Yip HK, Shao PL, Wu SC, Chen KH, Chen YT, Yang CC, Sun CK, Kao GS, Chen SY et al (2016) Combination of adipose-derived mesenchymal stem cells (ADMSC) and ADMSC-derived exosomes for protecting kidney from acute ischemia-reperfusion injury. Int J Cardiol 216:173–185

Lo Sicco C, Reverberi D, Balbi C, Ulivi V, Principi E, Pascucci L, Becherini P, Bosco MC, Varesio L, Franzin C et al (2017) Mesenchymal stem cell-derived extracellular vesicles as mediators of anti-inflammatory effects: endorsement of macrophage polarization. Stem Cells Transl Med 6:1018–1028

Lu Z, Wang G, Dunstan CR, Chen Y, Lu WY, Davies B, Zreiqat H (2013) Activation and promotion of adipose stem cells by tumour necrosis factor-alpha preconditioning for bone regeneration. J Cell Physiol 228:1737–1744

Ma J, Zhao Y, Sun L, Sun X, Zhao X, Sun X, Qian H, Xu W, Zhu W (2016) Exosomes derived from Akt-modified human umbilical cord mesenchymal stem cells improve cardiac regeneration and promote angiogenesis via activating platelet-derived growth factor D. Stem Cells Transl Med 6(1):51–59

Madrigal M, Rao KS, Riordan NH (2014) A review of therapeutic effects of mesenchymal stem cell secretions and induction of secretory modification by different culture methods. J Transl Med 12:260

Nassar W, El-Ansary M, Sabry D, Mostafa MA, Fayad T, Kotb E, Temraz M, Saad AN, Essa W, Adel H (2016) Umbilical cord mesenchymal stem cells derived extracellular vesicles can safely ameliorate the progression of chronic kidney diseases. Biomater Res 20:21

Nong K, Wang W, Niu X, Hu B, Ma C, Bai Y, Wu B, Wang Y, Ai K (2016) Hepatoprotective effect of exosomes from human-induced pluripotent stem cell-derived mesenchymal stromal cells against hepatic ischemia-reperfusion injury in rats. Cytotherapy 18:1548–1559

Ophelders DR, Wolfs TG, Jellema RK, Zwanenburg A, Andriessen P, Delhaas T, Ludwig AK, Radtke S, Peters V, Janssen L et al (2016) Mesenchymal stromal cell-derived extracellular vesicles protect the fetal brain after hypoxia-ischemia. Stem Cells Transl Med 5:754–763

Ramos T, Sánchez-Abarca LI, Muntión S, Preciado S, Puig N, López-Ruano G, Hernández-Hernández Á, Redondo A, Ortega R, Rodríguez C et al (2016) MSC surface markers (CD44, CD73, and CD90) can identify human MSC-derived extracellular vesicles by conventional flow cytometry. Cell Commun Signal 14:2

Ranghino A, Bruno S, Bussolati B, Moggio A, Dimuccio V, Tapparo M, Biancone L, Gontero P, Frea B, Camussi G (2017) The effects of glomerular and tubular renal progenitors and derived extracellular vesicles on recovery from acute kidney injury. Stem Cell Res Ther 8:24

Rani S, Ryan AE, Griffin MD, Ritter T (2015) Mesenchymal stem cell-derived extracellular vesicles: toward cell-free therapeutic applications. Mol Ther 23:812–823

Shao L, Zhang Y, Lan B, Wang J, Zhang Z, Zhang L, Xiao P, Meng Q, Geng YJ, Yu XY et al (2017) MiRNA-sequence indicates that mesenchymal stem cells and exosomes have similar mechanism to enhance cardiac repair. Biomed Res Int 2017:4150705

Tan CY, Lai RC, Wong W, Dan YY, Lim SK, Ho HK (2014) Mesenchymal stem cell-derived exosomes promote hepatic regeneration in drug-induced liver injury models. Stem Cell Res Ther 5:76

Teng X, Chen L, Chen W, Yang J, Yang Z, Shen Z (2015) Mesenchymal stem cell-derived exosomes improve the microenvironment of infarcted myocardium contributing to angiogenesis and anti-inflammation. Cell Physiol Biochem 37:2415–2424

Vallabhaneni KC, Penfornis P, Dhule S, Guillonneau F, Adams KV, Mo YY, Xu R, Liu Y, Watabe K, Vemuri MC et al (2015) Extracellular vesicles from bone marrow mesenchymal stem/stromal cells transport tumor regulatory microRNA, proteins, and metabolites. Oncotarget 6:4953–4967

Van Pham P, Truong NC, Le PT, Tran TD, Vu NB, Bui KH, Phan NK (2016a) Isolation and proliferation of umbilical cord tissue derived mesenchymal stem cells for clinical applications. Cell Tissue Bank 17:289–302

Van Pham P, Vu NB, Phan NK (2016b) Umbilical cord-derived stem cells (MODULATISTTM) show strong immunomodulation capacity compared to adipose tissue-derived or bone marrow-derived mesenchymal stem cells. Biomed Res Ther 3:687–696

Van Pham P, Vu NB, Phan NL-C, Le DM, Truong NC, Truong NH, Bui KH-T, Phan NK (2014) Good manufacturing practice-compliant isolation and culture of human adipose derived stem cells. Biomed Res Ther 1:21

Xing J, Hou T, Jin H, Luo F, Change Z, Li Z, Xie Z, Xu J (2014) Inflammatory microenvironment changes the secretory profile of mesenchymal stem cells to recruit mesenchymal stem cells. Cell Physiol Biochem 33:905–919

Yu B, Shao H, Su C, Jiang Y, Chen X, Bai L, Zhang Y, Li Q, Zhang X, Li X (2016) Exosomes derived from MSCs ameliorate retinal laser injury partially by inhibition of MCP-1. Sci Rep 6:34562

Zhang Y, Chopp M, Zhang ZG, Katakowski M, Xin H, Qu C, Ali M, Mahmood A, Xiong Y (2016) Systemic administration of cell-free exosomes generated by human bone marrow derived mesenchymal stem cells cultured under 2D and 3D conditions improves functional recovery in rats after traumatic brain injury. Neurochem Int 111:69–81

Zhao Y, Sun X, Cao W, Ma J, Sun L, Qian H, Zhu W, Xu W (2015) Exosomes derived from human umbilical cord mesenchymal stem cells relieve acute myocardial ischemic injury. Stem Cells Int 2015:761643

Zhou Y, Xu H, Xu W, Wang B, Wu H, Tao Y, Zhang B, Wang M, Mao F, Yan Y et al (2013) Exosomes released by human umbilical cord mesenchymal stem cells protect against cisplatin-induced renal oxidative stress and apoptosis in vivo and in vitro. Stem Cell Res Ther 4:34

Zou X, Gu D, Xing X, Cheng Z, Gong D, Zhang G, Zhu Y (2016) Human mesenchymal stromal cell-derived extracellular vesicles alleviate renal ischemic reperfusion injury and enhance angiogenesis in rats. Am J Transl Res 8:4289–4299

Zou X, Zhang G, Cheng Z, Yin D, Du T, Ju G, Miao S, Liu G, Lu M, Zhu Y (2014) Microvesicles derived from human Wharton's Jelly mesenchymal stromal cells ameliorate renal ischemia-reperfusion injury in rats by suppressing CX3CL1. Stem Cell Res Ther 5:40

Part II
Stem Cells

Chapter 6
Evolution of Stem Cell Products in Medicine: Future of Off-the-Shelf Products

Phuc Van Pham, Hoa Trong Nguyen, and Ngoc Bich Vu

Abbreviations

ADSC	Adipose-derived stem cells
BM	Bone marrow
CHMP	The Committee for Medicinal Products for Human Use
CSCC_ASC	Cryopreserved Cardiology Stem Cell Centre adipose-derived stromal cell
GMP	Good manufacturing practice
HLA	Human leukocyte antigen
IMDM	Iscove's Modified Dulbecco's Medium
MNC	Mononuclear cells
MSC	Mesenchymal stem cell
PRP	Platelet-rich plasma
SVF	Stromal vascular fractions
T2DM	Type 2 diabetes mellitus
UC	Umbilical cord

P. Van Pham (✉) · N. B. Vu
Laboratory of Stem Cell Research and Application, VNUHCM University of Science, Ho Chi Minh City, Vietnam

Stem Cell Institute, VNUHCM University of Science, Ho Chi Minh City, Vietnam
e-mail: pvphuc@hcmuns.edu.vn

H. T. Nguyen
Stem Cell Institute, VNUHCM University of Science, Ho Chi Minh City, Vietnam

P. V. Pham (ed.), *Stem Cell Drugs - A New Generation of Biopharmaceuticals*, Stem Cells in Clinical Applications, https://doi.org/10.1007/978-3-319-99328-7_6

Stem Cells and Stem Cell Therapy

Stem cells are unspecialized cells that can become specialized cells capable of performing particular functions (Bongso and Lee 2005). These cells determine human development as well as development of other mammals, from embryo to adult. The cells derived in blastocysts are known as embryonic stem cells; in adults, they are termed adult stem cells. Although there are some differences between embryonic stem cells and adult stem cells, they always display two main properties, which are self-renewal and differentiation potential. Self-renewal is the capacity of stem cells to undergo cell division for a long time while maintaining their stem cell properties; differentiation potential refers to the capacity of stem cells to become various functional cells.

Given these properties, stem cells can be used to replace failed or defective cells in the human body. The definition of "stem cell therapy" or "stem cell transplantation" refers to the use of stem cells in treatment or in medicine; additionally, the treatment of diseases by stem cells is referred to as "regenerative medicine". The application of stem cells in clinical applications has had a long history. The first transplantation of hematopoietic stem cells (HSCs) was used to treat leukemia in the 1950s by Dr. E. Donnall Thomas at the Fred Hutchinson Cancer Research Center in the USA (Thomas 2005). This treatment used stem cells from bone marrow from one identical twin to treat another. There were no problems with the transplantation because both twins shared the same genetics. In 1968, the second transplantation of HSCs in non-twins was performed (Antoine et al. 2003). This time, the key to a successful transplantation was determined by genetic matching (known as HLA matching) of the donor to the patient (Amos and Bach 1968).

Ten years later, HSCs were discovered in the human umbilical cord blood (Prindull et al. 1978). Following that, different kinds of stem cells, including embryonic stem cells (ESCs), mesenchymal stem cells (MSCs) (Friedenstein et al. 1976), and endothelial stem cells (EPCs) (Asahara et al. 1997), were discovered. To date, some thousands of clinical trials using stem cells have been performed in more than 20 countries to treat a variety of diseases and physiological conditions, such as retinal blindness, Parkinson's disease, Huntington's disease, spinal cord injury, myocardial infarction, and type II diabetes mellitus (Van Pham 2016a).

Generations of Stem Cell Products

The products of stem cells were first used in humans in the 1950s in the form of bone marrow (Thomas 2005). Since then, with more than 50 years of evolution, stem cell products have been used in the clinic and have shown significant change in both quality and formulation. The stem cell products are now greater in purity, or some have been enhanced in particular characteristics to increase their treatment efficacy.

Based on the purity level and characteristics of stem cells, nowadays stem cell products are often grouped into six generations:

- First generation: Stem cell-enriched fractions
- Second generation: Pure stem cells
- Third generation: Long-term expanded allogenic stem cells
- Fourth generation: Genetically modified or differentiated stem cells
- Fifth generation: Exosomes, extracellular vesicles, and stem cell extracts
- Sixth generation: Stem cells derived from tissues or organs.

Based on the differences in purity and characteristics of stem cells, the various generations have distinct advantages and disadvantages (Table 6.1).

First Generation: Products or Fractions of Stem Cell Enrichment

The stem cell-enriched product is produced primarily from two main tissues: blood and fat. The blood used for the production can be from bone marrow, peripheral blood (with mobilized stem cells), or cord blood. Adipose tissue can be obtained from thighs, buttocks, or belly (Table 6.2). The general feature of this product group is that the proportion of stem cells present in the mixture for transplants is relatively low, typically <10%. However, the advantages of this product group is its rapid production, minimal in vitro manipulation of cells, lower likelihood of undesirable mutations in the stem cells, and higher safety profile. Patients may be treated by autologous or allogenic stem cell transplants depending on disease conditions and stem cell quality.

However, this group of product also has some disadvantages. The main one is that the treatment efficacy is so different between patients due to difficulty in standardizing cell quality before grafting.

This group of products is widely used in many countries in the world where HSC-rich fractions are used primarily for the treatment of blood or genetic diseases (Fig. 6.1).

Second Generation: Pure Stem Cell Products

Pure stem cell products are a group of highly purified stem cell products that consist of >90% stem cells in the cell mixture; they are also used in transplant. Currently, the two main stem cell types in this group include HSCs and MSCs.

HSCs can be obtained from umbilical cord blood (Broxmeyer et al. 1989, 1991; Gluckman 2001), bone marrow (Alvarez et al. 2013; Kondo et al. 2003; Wilson and Trumpp 2006), and peripheral blood (Gluckman 2000; Sheridan et al. 1992; Carella

Table 6.1 Differences between various generations of therapeutic stem cells or stem cell products

Generation	Advantages	Disadvantages
Firstst generation: stem cell-enriched fraction	– Easy production – Minimal manipulation of stem cells such that the safety of product is high – Time for manufacturing is short – Cost for manufacturing is low	– Low stem cell purity and the number of stem cells is limited – Quality of product is difficult to control, and differs from batch to batch – Unstable treatment efficacy depending on the quality of the grafted sample – Difficult to scale up the manufacturing – Products are difficult to transport and store
Second generation: pure stem cells	– High purity of the stem cells, minimizing adverse effects from other contaminated cells – Cell quality may be partially controlled	– High production costs – Long production time – Difficult to scale up the manufacturing – Products are difficult to transport and store
Third generation: long-term expanded allogenic stem cells	– High stem cell purity – Quality of the product is strictly controlled – Can be produced by industrial scale – Cost of production is decreased due to large-scale production	– Requires a clean room BSL2 or higher – Investment in equipment for expansion and quality control – Spontaneous mutation or differentiation of stem cells can be carried out during the in vitro expansion
Fourth generation: genetically modified or differentiated stem cells	– High stem cell purity – Quality of the product is strictly controlled – Can be produced by industrial scale – Cost of production is decreased due to large-scale production – Stem cells can display some particular phenotypes or properties that can improve the treatment efficacy	– Requires a clean room BSL2 or higher – Investment in equipment for expansion and quality control – Spontaneous mutation or differentiation of stem cells can be carried out during the in vitro expansion – Stem cells "faked" or modulated by changes in genetics and epigenetics can increase the risk of tumorigenesis
Fifthth generation: exosomes, extracellular vesicles and stem cell extracts	– Product does not contain whole cells therefore immune response is low – Easy production by industrial scale – Product quality is easy to control – Easy to store and transport	– Multiple activities of stem cells are lost thus the regenerative effect of stem cells is changed or decreased compared to whole cells – Product quality control is relatively complex – High manufacturing costs

(continued)

Table 6.1 (continued)

Generation	Advantages	Disadvantages
Sixth generation: stem cells derived from tissues or organs	– Tissues or organs can directly replace defective tissues or organs so treatment efficacy can be rapidly improved – Effective treatment of tissue or organ defects/dysfunction	– Production technique is extremely complex – Difficult to scale up the processing – High production costs – Product quality difficult to control

Table 6.2 Some examples of first generation stem cell products

First generation stem cell products	Applications
Mononuclear cells (MNCs) from umbilical cord blood (enriched fraction of HSCs from umbilical cord blood)	Treatment of malignant diseases of blood, genetic disorders, heart diseases, etc.
MNCs from peripheral blood (enriched fraction of HSCs recruited from peripheral blood)	Treatment of malignant diseases of blood, genetic disorders, heart diseases, solid tumors, etc.
MNCs from bone marrow (enriched fraction of HSCs recruited from bone marrow)	Treatment of malignant diseases of blood, genetic disorders, heart diseases, solid tumors, osteoarthritis, brain disease, etc.
Stromal vascular fractions (SVFs) from adipose tissue	Osteoarthritis, cardiovascular disease, diabetes mellitus, chronic obstructive pulmonary disease, peptic ulcer disease, etc.

et al. 2000; Kang et al. 2004). The cells are purified through various purification techniques to obtain HSCs expressing the surface marker CD34 ($CD34^{+}$), or CD34 and CD45 ($CD34^{+}CD45^{+}$). Transplantation of purified products containing only CD34+ HSCs is thought to have many advantages (Negrin et al. 2000; Somlo et al. 1997; Lacerda et al. 2005; Oyekunle et al. 2006; Haen et al. 2015).

MSCs can also be obtained from umbilical cord blood (Van Pham et al. 2014a; Lee et al. 2004; Gang et al. 2004), bone marrow (Hung et al. 2002; Prockop et al. 2001; Mareschi et al. 2001), adipose tissue (Van Pham et al. 2014b; Li et al. 2018a, b), umbilical cord tissue (Van Pham et al. 2016a; Dehkordi et al. 2016), Wharton's jelly (Cardoso et al. 2012; Corotchi et al. 2013; Al Madhoun et al. 2016), and dental pulp (Poltavtseva et al. 2014; Hilkens et al. 2013; Mochizuki and Nakahara 2018). They are purified primarily through selective cell culture technology based on adherence of cells to the surface of culture vessels (Van Pham et al. 2014a, b, 2016a, b) or by sorting based on markers of MSCs (Hagmann et al. 2014; Kouroupis et al. 2014; Battula et al. 2009). Normally, these MSCs are highly purified after 3–5 passages of subculture and are ready to be used for transplantation (Van Pham et al. 2014a, 2016a, b). This group of products is widely used in many treatment facilities in the world. As part of the procedure, bone marrow, adipose tissue were collected from patients, would then be selectively cultured, followed by purification and proliferation of MSCs, and lastly by transplantation into the patient.

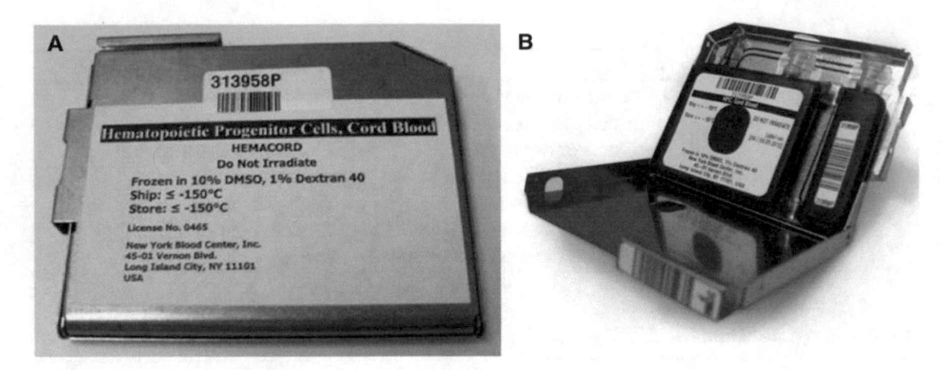

Fig. 6.1 Two products of first generation stem cells were approved in the USA. (**a**) Hemacord and (**b**) Ducord are products of enriched hematopoietic stem cells derived from umbilical cord blood

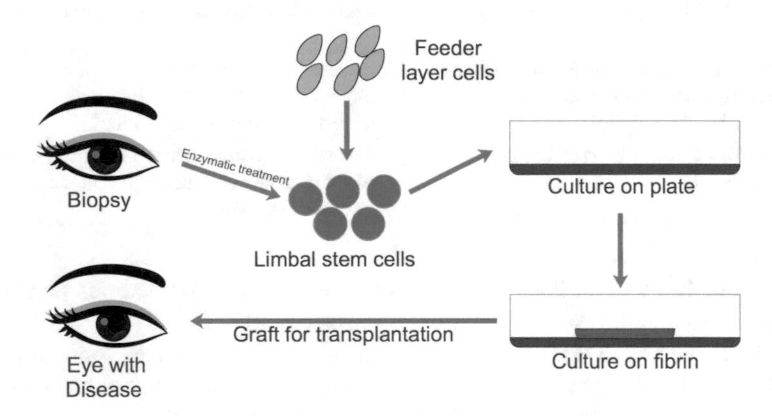

Fig. 6.2 The procedure of limbal stem cell manufacturing for clinical application. The above product (Holoclar) has been approved in Europe

Recently, Holoclar products have been licensed for circulation in Europe, and are an example of products of the second generation (Fig. 6.2). In the Holoclar products, stem cells from the limbal portion of the cornea (limbal stem cells) are cultured ex vivo and used in autologous transplantation (Pellegrini et al. 2018; Farkas et al. 2017).

Third Generation: Products of Stem Cells After Long-Term Expansion

This is a growing group of products which have opened up the multinational stem cell industry. The first product in this group, Prochymal, was manufactured in large-scale and approved for use in Canada in 2012 for the treatment of graft versus host disease (GVHD) (Vaes et al. 2012; Mannon 2011). So far, several similar products have been successfully produced and marketed in other countries; these products include Cartistem (Korea), HS TemCell (Japan), and Alofiscel (Europe) (Fig. 6.3).

Fourth Generation: Products of Modified or Differentiated Stem Cells

This is a generation of products consisting of stem cells which have been modified to enhance certain properties, exhibit novel properties, or differentiate into functional cells. This group is currently one of the most promising generations of stem cell products. Products from epigenetic reprogramming processes, which can reprogram certain cell phenotypes directly into functional cells, or from functional cells into stem cells, belong to this generation.

The most striking product of this generation is STRIMVELIS. This is a new product from GSK. This product is a fraction of CD34+ rich blood-forming HSCs encoded with adenosine deaminase (ADA) cDNA gene via a retroviral vector. It was licensed for circulation in Europe in 2016, and is used to treat severe immunodeficiency (SCID) due to ADA deficiency (Schimmer and Breazzano 2016; Monaco and Faccio 2017; Stirnadel-Farrant et al. 2018).

Recently, genetically modified transgenic T cells (chimeric antigen receptor T cells—Car-T) have also been considered as products of this generation.

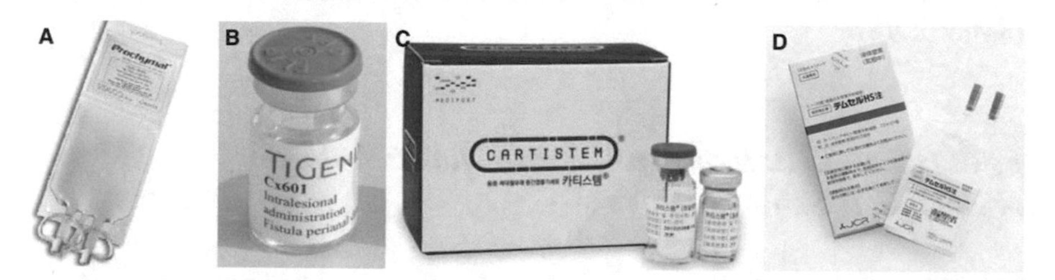

Fig. 6.3 Some third generation stem cell products that have been commercialized for use in several countries. (**a**) Prochymal (Cananda), (**b**) Alofis (Europe), (**c**) Cartistem (Korea), and (**d**) HS TemCell (Japan)

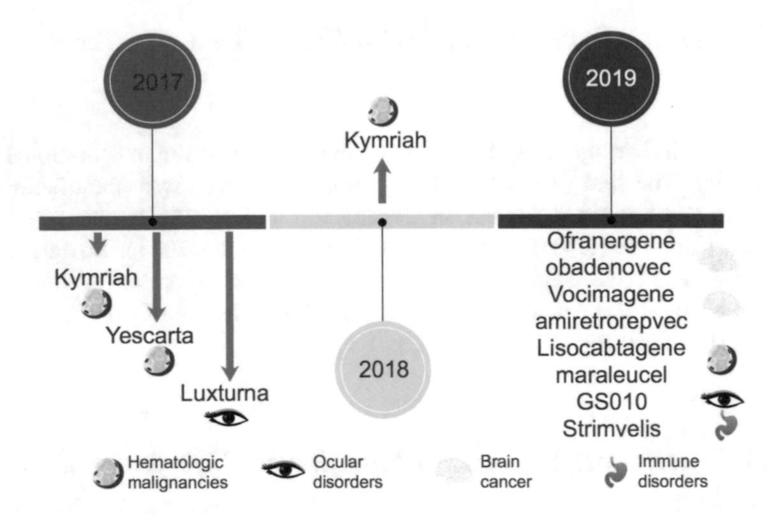

Fig. 6.4 Some genetically modified cell products applied in the treatment of various diseases

Transgenic products carrying mosaic receptors, such as Kymriah, Yescarta, and Luxturna, have been licensed for use in the treatment of certain cancers (Bach et al. 2017; Liu et al. 2017a; Silverman 2018) (Fig. 6.4). However, some scientists have argued that they are genetically modified cell products, not transgenic stem cell products.

Another line of stem cell products that has not been licensed for circulation in any country yet, but was supposed to pave the way for the future of stem cell products, is the product of induced pluripotent stem cells (iPSÇs). Notably, iPSCs can be produced by epigenetic reprogramming of any nucleated cells in the human body. The IPSCs can then be differentiated into specialized cells before transplantation in the body. The first-line treatment using these products in humans was conducted in Japan for the treatment of macular degeneration of the eye, in August 2013 (Mandai et al. 2017; Garber 2015; Reardon and Cyranoski 2014). Recently (since 2018), Japan has become the first country to report the clinical application and potency of iPSCs derived from cardiac muscle cells for the treatment of ischemic heart disease (Kyodo 2018).

Fifth Generation: Exosomes, Extracellular Vesicles or Stem Cell Extracts

The fifth generation of stem cell products includes stem cell secretome products, or stem cell-derived products produced during stem cell culture (e.g., extracellular vesicles, exosomes, cytokines, growth factors, enzymes, or even stem cell extracts). These components show many important biological activities (Fatima and Nawaz

Fig. 6.5 Various companies which have produced stem cells or stem cell progenitor cell products that have been approved and commercialized in the market

2015; Kim et al. 2018; Zhang et al. 2018; Kobayashi et al. 2018; Liu et al. 2017b). Clinical trials using such products are being initiated (Dehghani 2018; Giebel et al. 2017). Compared with products containing whole cells, stem cell secretome products have many advantages in production, including easy storage, packaging, and transport. However, the greatest difficulty in producing these products is the variation in stability and quality of the product across different batches. In fact, it has been found that the secretome of the same type of cell even varies at different times.

To date, some of the products of this generation have been manufactured and commercialized, such as Stemedica (Vitrilife), which contains stem-cell derived secretomes stored at room temperature. These products have been clinically tested on skin in 2016 (NCT01771679), and for heart failure in 2018.

Most applications of stem cell secretomes have been for cosmetic purposes. Products which have been commercialized as cosmetic products include Regencia cream (SkinMedica), Lifeline skin care (International Stem Cell Corporation), Blue Horizon, Celprogen, ReLuma Advanced Stem Cell Facial Moisturizer, ReLuma Skin Illuminating Stem Cell Anti-Aging Cleanser, and ReLuma Stem Cell Eye Cream, among others (Fig. 6.5).

Sixth Generation: Engineered Issues or Organs from Stem Cells

Engineered tissues or organs are special products of the stem cell industry, which is the culmination of three important industries, consisting of stem cells, biomaterials, and recombinant proteins. Indeed, a tissue or organ that is made to replace or treat a disease or defect in the body is the product of a combination of important components, namely stem cells, biological materials, and proteins.

Despite early promising results, stem cell-derived tissue or organ transplants are still low in terms of number and type. Most of these products are being produced individually by each patient for individual treatment. Indeed, after induction into specialized cells in tissues, these cells strongly express HLA surface antigens which increase immunogenicity and are readily excluded when transplanted into patients.

Generations of Stem Cell Products and Technology Requirements

Generally, stem cell products from the different generations require different technologies. The evolution of stem cell products are synchronized with the evolution of electronic, automatic, and informatic innovations. Indeed, the first generation of stem cell products have required simple techniques to enrich the stem cells and partly remove the undesired cells, such as red blood cells. However, in the later generations, the manufacturing of stem cell products has required more complex, modern techniques to purify, expand, modify, differentiate, and store the stem cells.

Stem cell technology now includes four core technologies: (1) isolation or enrichment technology; (2) proliferation or expansion; (3) modification or differentiation; and (4) storage or cryopreservation (Table 6.3).

Isolation and Enrichment

Depending on whether the tissues are solid or liquid, the techniques used to isolate or enrich the stem cells are different. Some liquid tissues include peripheral blood, bone marrow, menstrual blood, milk, umblical cord blood, and amniotic liquid. These tissues can be used directly to enrich stem cells using some simple techniques before they are purified.

Depending on the kind of stem cells, the cells can be enriched by centrifugation or via in vitro culture. Generally, HSCs can be enriched from bone marrow, peripheral blood, and umbilical cord blood by Ficoll gradient centrifugation. Meanwhile, MSCs can be enriched by in vitro culture and will adhere onto the surface of culture vessels.

Table 6.3 Some technologies for the generation of stem cell products

Generation of stem cell products	Core technologies used in the manufacturing process
First generation: stem cells enriched fraction	– Enzymatic isolation for solid tissues, for example collagenase, lecithin – Gradient centrifugation – Elutriation centrifugation – Filter
Second generation: pure stem cells	– Enzymatic isolation for solid tissues, for example collagenase, lecithin – Gradient centrifugation – Elutriation centrifugation – Filter – Magnetic-activated cell sorting – Fluorescence-activated cell sorting – Stem cell culture/tissue culture expansion
Third generation: long-term expanded allogenic stem cells	– Enzymatic isolation for solid tissues, for example collagenase, lecithin – Gradient centrifugation – Elutriation centrifugation – Filter – Magnetic-activated cell sorting – Fluorescence-activated cell sorting – Stem cell culture/tissue culture expansion – Scale-up stem cell culture (in bioreactors) – Xeno free, serum free stem cell culture – Cryopreservation of stem cells
Fourth generation: genetically modified or differentiated stem cells	– Enzymatic isolation for solid tissues, for example collagenase, lecithin – Gradient centrifugation – Elutriation centrifugation – Filter – Magnetic-activated cell sorting – Fluorescence-activated cell sorting – Stem cell culture/tissue culture expansion – Scale-up stem cell culture (in bioreactors) – Xeno free, serum free stem cell culture – Cryopreservation of stem cells – In vitro differentiation of stem cells – In vitro genetic modification of stem cells – Epigenetic reprogramming – Direct epigenetic reprogramming

(continued)

Table 6.3 (continued)

Generation of stem cell products	Core technologies used in the manufacturing process
Fifth generation: exosomes, extracellular vesicles and stem cell extracts	– Enzymatic isolation for solid tissues, for example collagenase, lecithin – Gradient centrifugation – Elutriation centrifugation – Filter – Magnetic-activated cell sorting – Fluorescence-activated cell sorting – Stem cell culture/tissue culture expansion – Scale-up stem cell culture (in bioreactors) – Xeno free, serum-free stem cell culture – Cryopreservation of stem cells – In vitro differentiation of stem cells – In vitro genetic modification of stem cells – Epigenetic reprogramming – Direct epigenetic reprogramming – Freeze drying – Exosome, extracellular vesicle isolation, extraction – Ultracentrifugation – Stem cell extract process
Sixth generation: stem cells derived tissues or organs	– Enzymatic isolation for solid tissues, for example collagenase, lecithin – Gradient centrifugation – Elutriation centrifugation – Filter – Magnetic-activated cell sorting – Fluorescence-activated cell sorting – Stem cell culture/tissue culture expansion – Scale-up stem cell culture (in bioreactors) – Xeno free, serum-free stem cell culture – Cryopreservation of stem cells – In vitro differentiation of stem cells – In vitro genetic modification of stem cells – Epigenetic reprogramming – Direct epigenetic reprogramming – Biomaterials/scaffolds – Growth factors – 3D stem cell culture and differentiation

Stem cell-enriched fractions can be purified by techniques, such as magnetic-activated cell sorting (MACS), electric fields in fluorescence-activated cell sorting (FACS), tube catchers, microfluid chips, immune panning, laser capture microdissection, and selection under microscopy.

For solid tissues, the stem cells can be isolated by in vitro culture, or enrichment and purification similar to fluid tissues, after they are dissocciated into single cells. Unlike adult cells in tissues, stem cells can undergo self-renewal for a long period and can proliferate long-term during in vitro cell culture. These characteristics have allowed them to be isolated after in vitro cell culture. Using this technique, the tissue can be fragmented into tiny fragments and then cultured for tissue expansion. After about 1–2 weeks, stem cells and some progenitor cells will migrate out from the tissue and adhere to the surface of the culture vessels. In another method, the tissue is dissociated into single cells and the suspension is then cultured in vitro. The progenitor stem cells in the suspension then adhere to the surface of the culture vessels and proliferate.

Proliferation and Expansion

There are important technologies to expand stem cells. The stem cells usually exist at a minimal quantity in tissues to maintain tissue homeostasis, replacing aged or injured cells via the process of self-renewal and differentiation. For stem cell therapy, the stem cells should be expanded to adequately high numbers. However, the most important consideration to keep in mind is the maintenance of stem cell phenotype and functional stem cell properties during the expansion process.

Stem cells can be expanded by either adherent or suspension culture conditions. In the adherent culture condition, stem cells will attach onto the suface of the culture vessels and proliferate; in the suspension culture, stem cells are free to proliferate in the culture medium. Depending on the kind of stem cells as well as the culture technologies, one of these platforms is utilized to expand stem cells in vitro.

In the adherent cell culture, stem cells can be proliferated in 2D culture (i.e., in monolayer form) or in 3D culture (in 3D form such as spheres or pellets). Some studies have suggested that 3D culture helps to maintain and enhance certain properties of stem cells, especially MSCs, compared to 2D culture. To carry out 3D culture, stem cells can be cultured in specialized media or conditions that allow the formation of pellets or spheres (Cesarz and Tamama 2016; Li et al. 2015). For examples, culture of stem cells in hanging drop plates allows for this 3D growth (Schmal et al. 2016). Another method of 3D culture is bead culture. In this technique, stem cells are seeded on the surface of microbeads, which are then cultured in a bioreactor or spinner flask (Dias et al. 2017; Hervy et al. 2014). Recently, a new technology invented by Terumo permits the scale up of adherent culture in vitro via the use of hollow flow fibers (Quantum Cell expansion system) (Rojewski et al. 2013; Roberts et al. 2012). In this system, stem cells can adhere onto fibers with about 2.1 m^2 cell culture surface area.

Conversely, in the cell suspension culture, stem cells are free in the culture media or can attach to beads that are free-floating in the culture media. These free cells or beads with cells are then placed in bioreactors controlled by a stirring apparatus, or in wave bioreactors. Most stem cells need to attach to a surface to proliferate, except for HSCs. Therefore, the cell suspension culture for free cells in media is best suited for expansion of HSCs, but not for expansion of adherent cells, such as MSCs and iPSCs.

Modification and Differentiation

Modification and differentiation are two processes of engineering which modify or convert stem cells into specialized stem cells (or specialized cells). Stem cells can perform more functions beyond their stemness properties. For example, stem cells can be made to overexpress certain proteins which help trigger the wound healing process, or can be used as vectors to deliver therapeutic proteins to target tissues or organs. Indeed, HSCs have been modified to express certain enzymes to treat genetic disorders, and MSCs have been modified to express enzymes and factors to treat cancers (Schimmer and Breazzano 2016).

Differentiation is the process by which stem cells achieve particular functions after significant epigenetic changes. In some cases, stem cells are differentiated into functional cells before they can be used in therapy (Mandai et al. 2017). Some methods which have been used to differentiate stem cells include chemical methods, physical agents, and biological agents. Certain chemicals have been demonstrated to be differentiating agents that can cause the epigenetic changes toward functional cells. For example, 5-aza-cytidine can cause the differentiation of MSCs to cardiomyocyte-like cells (Bae et al. 2017; Wan Safwani et al. 2012; Qian et al. 2012), and dexamethasone can cause the differentiation of MSCs to osteoblasts (Chen et al. 2016; Ghali et al. 2015). Moreover, certain physical conditions, including hypoxia and light, can also trigger some physiological processes of stem cells (Boyette et al. 2014; Lan et al. 2015; Yuan et al. 2017; Li et al. 2010).

Biological agents are popular factors to differentiate stem cells. These agents can be growth factors, vesicles or exosomes from other cells, coculture conditions, tissue or cell extraction factors, genes (DNA, mRNA, etc.), or small molecules (miRNA, siRNA, shRNA, etc.) (Akiyama et al. 2018; Xie et al. 2007; Mehta et al. 2014). By using these factors, most stem cells can be differentiated into functional cells. Moreover, these factors can also induce epigenetic reprogramming of cells or stem cells; in other words, the factors help to achieve the desired cellular phenotypes from differentiated cells. Indeed, epigenetic reprogramming is an important technology, and was firstly demonstrated as a method to produce iPSCs by Yamanaka et al. (Takahashi and Yamanaka 2006).

Storage or Cryopreservation

Storage or conservation is the final step of stem cell manufaturing. This process helps to maintain the stem cell products for later use. There are two groups of technologies used to store stem cells, namely cyropreservation and dry freezing. While cryopreservation of stem cell products has been developed and applied for a long time, the dry freezing method is a relatively newer method which has been recently studied.

In cryopreservation, stem cell products are mixed in the cryopreservation media with cold stress protectants. Based on the rate of cooling, the cryopreservation techniques can be grouped into three kinds: rapid freezing, programmed freezing, and vitrification. Most stem cells, including HSCs, can be cryopreserved by programmed freezing (Winter et al. 2014; Reich-Slotky et al. 2008; Devadas et al. 2017). However, vitrification has only been recently studied for the cryopreservation of MSCs (Zanata et al. 2018; Fu et al. 2017; Massood et al. 2013; Bhakta et al. 2009).

The cryopreservation media is also an important component which helps to increase the percent viability of cells after thawing. There are various formulations (generations) of culture media, including media containing bovine serum, human serum/platelet-rich plasma (Wang et al. 2017), and/or defined chemicals (Miyagi-Shiohira et al. 2017; Lauterboeck et al. 2016; Roy et al. 2014). Some media differ in their cooling protectants and some contain DMSO, while others contain DMSO in combination with glycerol. The more recent media formulations are free of DMSO and glycerol (Wang et al. 2011; Rodrigues et al. 2008; Miyamoto et al. 2012; Rogulska et al. 2017; Dovgan et al. 2017; Shivakumar et al. 2016).

Treatment Efficacy of Various Generations of Stem Cell Products

First Generation Stem Cell Products Compared to Second Generation Stem Cell Products

The treatment efficacy of different generations of stem cell products was studied and reported for various diseases. Most reports compared the treatment efficacy of first generation stem cell products with second generation stem cell products.

Some studies performed comparisons of stromal vascular fractions (SVFs) from adipose tissue (i.e., first generation stem cell product) with adipose-derived stem cells (ADSCs) (i.e., second generation stem cell product) in both preclinical trials (in animals) and clinical trials (in humans). The study by Domergue et al. (Domergue et al. 2016) compared the efficacy of SVFs and ADSCs in hypertrophic scar (HTS) treatment in nude mice. Their results showed that although both SVFs and ADSCs could attenuate the HTS, ADSCs appeared more effective than SVFs. Transplantation of modified ADSCs could confer expression of transforming growth factor-beta (TGF-β3) and hepatic growth factor (HGF) (Domergue et al. 2016).

However, in a rat model of acute renal ischemia–reperfusion injury, L. Zhou et al. showed equivalent treatment efficacy of ADSCs and SVFs in attenuating acute renal IR injury. Both SVFs and ADSCs were transplanted into injured kidney through intraparenchymal injection. Renal functions for both treated groups were significantly improved, and there was reduced tubular injury, improved cell proliferation, and markedly reduced cellular apoptosis (Zhou et al. 2017).

In a rat model of hind limb ischemia, Iwase et al. showed that MSC transplantation significantly improved hind limb ischemia compared to MNC transplantation. Indeed, the MSC transplantation induced greater capillary density compared to MNC transplantation or placebo. In particular, the vascular smooth muscle cells formed from the transplanted cells were detected in the MSC-transplanted group (Iwase et al. 2005). MSC transplantation was also better than MNC transplantation for myocardial infarction treatment in a rat model. In a study by Mazo et al., transplanted MSCs and MNCs obtained from bone marrow were compared for their ability to treat myocardial infarction. Similar to other studies, the results of their study showed that both MSCs and MNCs induced therapeutic effects in rats, but only MSCs could improve metabolism, which was accompanied with smaller infarct size, scar collagen content, and higher revascularization degree (Mazo et al. 2010).

MSC transplantation, too, has yielded better results than MNC transplantation in diabetic critical limb ischemia and foot ulcer treatment (Lu et al. 2011). In a pilot trial, Lu et al. (2011) evaluated 41 type 2 diabetes mellitus (T2DM) patients with bilateral critical limb ischemia and foot ulcer. The patients were randomly divided into three groups: MSC transplantation, MNC transplantation, and normal saline. After 6 weeks following transplantation, the ulcer healing rate was highest in the MSC transplantation group, compared with the MNC and normal saline groups (Lu et al. 2011). Moreover, a recent clinical trial compared MSCs versus MNCs from bone marrow in the treatment of T2DM (Bhansali et al. 2017). In that study, 30 patients with T2DM were randomized to receive bone marrow derived MSCs (BM-MSCs) or bone marrow derived MNCs (BM-MNCs), or a sham ($n = 10$ per group). The results showed that both infusion of BM-MSCs and BM-MNCs resulted in sustained reduction of insulin doses in the T2DM patients. Notably, there was an improvement in insulin sensitivity in the MSC transplantation group, while there was an increase in C-peptide production in the MNC transplantation group (Bhansali et al. 2017).

The first generation stem cell product of HSCs (i.e., MNCs enriched with CD34+ cells) was also compared to the second generation stem cell product of HSCs (i.e., pure CD34+ cells) in several studies. In a study of cerebral palsy treatment, umbilical cord blood derived MNCs and CD34+ cells were used; 30 mice were treated with MNCs and 30 were treated with CD34+ cells. Treatment with MNCs or CD34+ cells suppressed apoptotic gene expression and restored memory and motor function. However, the results indicated that CD34+ cell transplantation was significantly better at treating mice with cerebral palsy (Li et al. 2014)

In another study, CD34+ cells also showed superior efficacy, over unselected circulating MNCs, in preserving myocardial integrity and function after myocardial infarction in nude rats (Kawamoto et al. 2006). In this study by Kawamoto et al., human CD34+ cells from peripheral blood and MNCs (also from peripheral blood) were evaluated. The CD34+ cells were isolated from MNCs by magnetic cell sorting. Then, athymic nude rats were intramyocardially transplanted with 5×10^5 CD34+ cells/kg, 5×10^5 MNCs/kg, or a higher dose of MNCs (hiMNCs) containing 5×10^5 CD34+ cells/kg. The results confirmed that echocardiographic regional wall motion score was better preserved in the CD34+ cell group (21.8 ± 0.5) than in the PBS, loMNC, or hiMNC groups (Kawamoto et al. 2006). CD34+ cells were also confirmed to be more effective than MNCs in nonunion healing following bone fracture. The authors showed that there was similar augmentation of blood flow recovery at peri-nonunion sites in both CD34+ cell- and MNC-transplantation groups. However, a superior effect on nonunion repair, as demonstrated by radiological, histological, and functional assessment, was observed in the CD34+ cell group as compared to the MNC group (Fukui et al. 2015).

The Second Generation Stem Cell Products Compared to Fourth Generation Stem Cell Products

The fourth generation stem cell products are genetically modified stem cells or differentiated cells that can exhibit particular properties. Choi et al. (2016) transduced cytotoxic T-lymphocyte-associated protein 4 immunoglobulin (CTLA4Ig) in ADSCs and produced a cell line from CTLA4Ig-expressing ADSCs (termed CTLA4Ig-ADSCs). Next, Choi et al. used this cell line to treat sustained severe rheumatoid arthritis, and compared its efficacy to that of original ADSCs. In a sustained severe rheumatoid arthritis mouse model, CTLA4Ig-ADSC transplantation was more effective than ADSC transplantation. Type 2 collagen (CII) autoantibodies and C-terminal telopeptide of CII were both significantly decreased in the CTLA4Ig-ADSC transplantation group (Choi et al. 2016).

The MSCs transduced to express neurotrophin MNTS1 (a multineurotrophin that binds TrkA, TrkB, and TrkC), and p75(NTR) receptors or MSC-MNTS1/p75(−) (which bind mainly to the Trk receptors), also showed reduced inflammation and cystic cavity size, compared to control rats. Interestingly, only the transduced MSCs enhanced axonal growth and significantly prevented cutaneous hypersensitivity after spinal cord injury (Kumagai et al. 2013).

In another study, Xue et al. (2015) modified MSCs with the Bcl-xL gene. Then, the MSCs expressing Bcl-xL were used to treat heart infarction, and their efficacy was compared to that of unmodified MSCs. The authors showed that the Bcl-xL-transduced MSCs were more effective than wild-type MSCs, and that cell apoptosis was significantly decreased by Bcl-xL-transduced MSCs compared to wild-type MSCs (40% vs 26%, respectively) (Xue et al. 2015).

To enhance the immune modulation of ADSCs, Liu et al. (Liu et al. 2017c) transfected ADSCs with OX40-Ig fusion protein (OX40Ig) to create OX40Ig-expressing ADSCs. These cells were then transplanted into a Lewis rat model of renal allograft, with the aim of evaluating and validating their immunosuppressive activity compared to unmodified ADSCs. Although the results showed that both OX40Ig-ADSCs and ADSCs could significantly suppress T cell proliferation and increase the percentage of CD4+CD25+ regulatory T-cells, evidently the OX40Ig-ADSCs were more effective. The administration of OX40Ig-ADSCs markedly prolonged the mean survival time of renal grafts, reduced allograft rejection, downregulated the expression of intragraft interferon-gamma (IFN-γ), and upregulated the expression of several genes, including interleukin (IL)-10, TGF-β, and forkhead box protein 3 (Foxp3) (Liu et al. 2017c).

In a recent study, Kayoko Yanagihara et al. (2018) showed that MSCs expressing Runx2 (by transfection) and cultured in 3D platform (spheroids) could significantly induce bone regeneration, compared to non-transfected MSC spheroids.

The Treatment Efficacy of Third Generation Stem Cell Products (Off-the-Shelf Stem Cells)

Unlike the first, second, and fourth generation stem cell products, and most stem cells used in transplantation (i.e., autologous cells), the third generation stem cell products are allogenic stem cells. These cells were expanded in vitro for a long period to generate high enough numbers of stem cells; then they are stored into doses ready for future use. As shown in Fig. 6.3, to date there are some approved stem cell products of this generation that have been successfully commercialized in some countries. Such products include Prochymal (Cananda), Alofis (Europe), Cartistem (Korea), and HS TemCell (Japan). Other products are currently being investigated as well (Van Pham et al. 2016b; Kastrup et al. 2017; Le et al. 2016; Van Pham 2016b).

The commercialized products have shown potential treatment efficacy for certain diseases. For example, Prochymal was used to treat GVHD with promising results (Locatelli et al. 2017; Chen et al. 2014; Kurtzberg et al. 2014; Prasad et al. 2011), and Alofis (Cx601) was used to treat Crohn's disease with safe and effective results (Panes et al. 2018).

In a clinical trial study, the product CSCC_ASC (cryopreserved Cardiology Stem Cell Centre adipose-derived stromal cell)—i.e., allogenic stem cells from adipose tissue—was used to treat ten patients with ischemic heart disease and ischemic heart failure. The results showed that treatment with CSCC_ASC could improve cardiac function at 6-month follow-up, that left ventricular end systolic volume decreased, and that left ventricular ejection fraction increased (Kastrup et al. 2017).

The off-the-shelf stem cell products of neural stem cells have also been developed and evaluated for nerve tissue repair in animals. An example of such products is CTX0E03, a conditionally immortalized human neural stem cell line. The cells were used to produce engineered neural tissue (EngNT-CTX) and used to repair a 12 mm sciatic nerve injury model in athymic nude rats. The use of EngNT-CTX supported growth of neurites and vasculature at the injury site (O'Rourke et al. 2018).

Perpectives and Conclusion

Stem cell products have evolved from first to sixth generation to adapt to the demands of patients. Although the early generation of stem cell products showed some therapeutic effacy in patients, there are still limitations associated with those products. The greatest limitation of first and second generation stem cell products is that it is difficult to control the quality of the products before they can be used for patient treatment. The quality as well as characteristics of stem cells vary greatly from patient to patient, and between young and old patients. Therefore, autologous stem cell transplantation was a way to improve therapeutic efficacy, although its efficiency also varies from patient to patient. The third generation stem cell products (i.e., off-the-shelf stem cells) are a new generation of stem cell products which can theoretically resolve the limitations of first and second generation stem cell products.

Indeed, by careful selection of donors whose tissues are used to extract the stem cells, the quality of the product has been partly improved. Moreover, the products are produced according to GMP-compliant guidelines and, thus, the quality of the final product should be improved during the manufacturing process.

The fourth generation stem cell products have shown initial promising results in animals. However, there have been many indications of genetic instability during their genetic modifications. Thus, development of fourth generation stem cell products will require more studies in order to fully evaluate their safety before they can be used in humans.

The fifth generation of stem cell products consists of acellular products of stem cells, such as extracellular vesicles, exosomes or stem cell extracts. Although these products have shown promising results in both animals and humans, their very complex manufacturing process, as well as the disappearance of some cellular properties of stem cells, has suggested that they are more suitable as add-on therapy rather than main therapy for diseases.

The sixth generation is the future generation of stem cell products. With these products, a patient's tissues or organs can be directly replaced. However, the in vitro organogenesis of stem cells is still not clearly understood. To date, stem cell scientists have not controlled this process well, nor have they understood the production of completed tissues or organs with all their functions and properties.

Taken together, from all the analysis in this review, we believe that the third generation of stem cell products (i.e., off-the-shelf stem cells) is the most suitable generation of stem cell products, and should be a predominant focus of study and further development.

Acknowledgment This research was partly funded by Ministry of Science and Technology, Vietnam under grant number DM.10.DA/15; by Fostering Innovation through Research, Science and Technology via project 15/FIRST/2a/SCI.

References

Akiyama T, Sato S, Chikazawa-Nohtomi N, Soma A, Kimura H, Wakabayashi S et al (2018) Efficient differentiation of human pluripotent stem cells into skeletal muscle cells by combining RNA-based MYOD1-expression and POU5F1-silencing. Sci Rep 8(1):1189

Al Madhoun A, Ali H, AlKandari S, Atizado VL, Akhter N, Al-Mulla F et al (2016) Defined three-dimensional culture conditions mediate efficient induction of definitive endoderm lineage from human umbilical cord Wharton's jelly mesenchymal stem cells. Stem Cell Res Ther 7(1):165

Alvarez P, Carrillo E, Velez C, Hita-Contreras F, Martinez-Amat A, Rodriguez-Serrano F et al (2013) Regulatory systems in bone marrow for hematopoietic stem/progenitor cells mobilization and homing. Biomed Res Int 2013:312656

Amos DB, Bach FH (1968) Phenotypic expressions of the major histocompatibility locus in man (HL-A): leukocyte antigens and mixed leukocyte culture reactivity. J Exp Med 128(4):623–637

Antoine C, Muller S, Cant A, Cavazzana-Calvo M, Veys P, Vossen J et al (2003) Long-term survival and transplantation of haemopoietic stem cells for immunodeficiencies: report of the European experience 1968–99. Lancet (Lond) 361(9357):553–560

Asahara T, Murohara T, Sullivan A, Silver M, van der Zee R, Li T et al (1997) Isolation of putative progenitor endothelial cells for angiogenesis. Science 275(5302):964–967

Bach PB, Giralt SA, Saltz LB (2017) FDA approval of tisagenlecleucel: promise and complexities of a $475000 cancer drug. JAMA 318(19):1861–1862

Bae Y-J, Kwon Y-R, Kim HJ, Lee S, Kim Y-J (2017) Enhanced differentiation of mesenchymal stromal cells by three-dimensional culture and azacitidine. Blood Res 52(1):18–24

Battula VL, Treml S, Bareiss PM, Gieseke F, Roelofs H, de Zwart P et al (2009) Isolation of functionally distinct mesenchymal stem cell subsets using antibodies against CD56, CD271, and mesenchymal stem cell antigen-1. Haematologica 94(2):173–184

Bhakta G, Lee KH, Magalhaes R, Wen F, Gouk SS, Hutmacher DW et al (2009) Cryopreservation of alginate-fibrin beads involving bone marrow derived mesenchymal stromal cells by vitrification. Biomaterials 30(3):336–343

Bhansali S, Dutta P, Kumar V, Yadav MK, Jain A, Mudaliar S et al (2017) Efficacy of autologous bone marrow-derived mesenchymal stem cell and mononuclear cell transplantation in type 2 diabetes mellitus: a randomized, placebo-controlled comparative study. Stem Cells Dev 26(7):471–481

Bongso A, Lee EH (2005) Stem cells: from bench to bedside. World Scientific, Singapore

Boyette LB, Creasey OA, Guzik L, Lozito T, Tuan RS (2014) Human bone marrow-derived mesenchymal stem cells display enhanced clonogenicity but impaired differentiation with hypoxic preconditioning. Stem Cells Transl Med 3(2):241–254

Broxmeyer HE, Douglas GW, Hangoc G, Cooper S, Bard J, English D et al (1989) Human umbilical cord blood as a potential source of transplantable hematopoietic stem/progenitor cells. Proc Natl Acad Sci 86(10):3828–3832

Broxmeyer H, Kurtzberg J, Gluckman E, Auerbach A, Douglas G, Cooper S et al (1991) Umbilical cord blood hematopoietic stem and repopulating cells in human clinical transplantation. Blood Cells 17(2):313–329

Cardoso TC, Ferrari HF, Garcia AF, Novais JB, Silva-Frade C, Ferrarezi MC et al (2012) Isolation and characterization of Wharton's jelly-derived multipotent mesenchymal stromal cells obtained from bovine umbilical cord and maintained in a defined serum-free three-dimensional system. BMC Biotechnol 12:18

Carella AM, Cavaliere M, Lerma E, Ferrara R, Tedeschi L, Romanelli A et al (2000) Autografting followed by nonmyeloablative immunosuppressive chemotherapy and allogeneic peripheral-blood hematopoietic stem-cell transplantation as treatment of resistant Hodgkin's disease and non-Hodgkin's lymphoma. J Clin Oncol 18(23):3918–3924

Cesarz Z, Tamama K (2016) Spheroid culture of mesenchymal stem cells. Stem Cells Int 2016:9176357

Chen GL, Paplham P, McCarthy PL (2014) Remestemcel-L for acute graft-versus-host disease therapy. Expert Opin Biol Ther 14(2):261–269

Chen Q, Shou P, Zheng C, Jiang M, Cao G, Yang Q et al (2016) Fate decision of mesenchymal stem cells: adipocytes or osteoblasts? Cell Death Differ 23(7):1128–1139

Choi EW, Shin IS, Song JW, Lee M, Yun TW, Yang J et al (2016) Effects of transplantation of CTLA4Ig-overexpressing adipose tissue-derived mesenchymal stem cells in mice with sustained severe rheumatoid arthritis. Cell Transplant 25(2):243–259

Corotchi MC, Popa MA, Remes A, Sima LE, Gussi I, Lupu PM (2013) Isolation method and xeno-free culture conditions influence multipotent differentiation capacity of human Wharton's jelly-derived mesenchymal stem cells. Stem Cell Res Ther 4(4):81

Dehghani L (2018) Allogenic mesenchymal stem cell derived exosome in patients with acute ischemic stroke. https://clinicaltrials.gov/ct2/show/NCT03384433

Dehkordi MB, Madjd Z, Chaleshtori MH, Meshkani R, Nikfarjam L, Kajbafzadeh AM (2016) A simple, rapid, and efficient method for isolating mesenchymal stem cells from the entire umbilical cord. Cell Transplant 25(7):1287–1297

Devadas SK, Khairnar M, Hiregoudar SS, Ojha S, Punatar S, Gupta A et al (2017) Is long term storage of cryopreserved stem cells for hematopoietic stem cell transplantation a worthwhile exercise in developing countries? Blood Res 52(4):307–310

Dias AD, Elicson JM, Murphy WL (2017) Microcarriers with synthetic hydrogel surfaces for stem cell expansion. Adv Healthc Mater 6(16):PMID 28509413

Domergue S, Bony C, Maumus M, Toupet K, Frouin E, Rigau V et al (2016) Comparison between stromal vascular fraction and adipose mesenchymal stem cells in remodeling hypertrophic scars. PLoS One 11(5):e0156161

Dovgan B, Barlic A, Knezevic M, Miklavcic D (2017) Cryopreservation of human adipose-derived stem cells in combination with trehalose and reversible electroporation. J Membr Biol 250(1):1–9

Farkas AM, Mariz S, Stoyanova-Beninska V, Celis P, Vamvakas S, Larsson K et al (2017) Advanced therapy medicinal products for rare diseases: state of play of incentives supporting development in Europe. Front Med 4:53

Fatima F, Nawaz M (2015) Stem cell-derived exosomes: roles in stromal remodeling, tumor progression, and cancer immunotherapy. Chin J Cancer 34(12):541–553

Friedenstein AJ, Gorskaja JF, Kulagina NN (1976) Fibroblast precursors in normal and irradiated mouse hematopoietic organs. Exp Hematol 4(5):267–274

Fu X, Yan Y, Li S, Wang J, Jiang B, Wang H et al (2017) Vitrification of rhesus macaque mesenchymal stem cells and the effects on global gene expression. Stem Cells Int 2017:3893691

Fukui T, Mifune Y, Matsumoto T, Shoji T, Kawakami Y, Kawamoto A et al (2015) Superior potential of CD34-positive cells compared to total mononuclear cells for healing of nonunion following bone fracture. Cell Transplant 24(7):1379–1393

Gang EJ, Hong SH, Jeong JA, Hwang SH, Kim SW, Yang IH et al (2004) In vitro mesengenic potential of human umbilical cord blood-derived mesenchymal stem cells. Biochem Biophys Res Commun 321(1):102–108

Garber K (2015) RIKEN suspends first clinical trial involving induced pluripotent stem cells. Nat Biotechnol 33(9):890–891

Ghali O, Broux O, Falgayrac G, Haren N, van Leeuwen JPTM, Penel G et al (2015) Dexamethasone in osteogenic medium strongly induces adipocyte differentiation of mouse bone marrow stromal cells and increases osteoblast differentiation. BMC Cell Biol 16:9

Giebel B, Kordelas L, Borger V (2017) Clinical potential of mesenchymal stem/stromal cell-derived extracellular vesicles. Stem Cell Investig 4:84

Gluckman E (2000) Current status of umbilical cord blood hematopoietic stem cell transplantation. Exp Hematol 28(11):1197–1205

Gluckman E (2001) Hematopoietic stem-cell transplants using umbilical-cord blood. N Engl J Med 344(24):1860–1861

Haen SP, Schumm M, Faul C, Kanz L, Bethge WA, Vogel W (2015) Poor graft function can be durably and safely improved by CD34+–selected stem cell boosts after allogeneic unrelated matched or mismatched hematopoietic cell transplantation. J Cancer Res Clin Oncol 141(12):2241–2251

Hagmann S, Frank S, Gotterbarm T, Dreher T, Eckstein V, Moradi B (2014) Fluorescence activated enrichment of CD146+ cells during expansion of human bone-marrow derived mesenchymal stromal cells augments proliferation and GAG/DNA content in chondrogenic media. BMC Musculoskelet Disord 15:322

Hervy M, Weber JL, Pecheul M, Dolley-Sonneville P, Henry D, Zhou Y et al (2014) Long term expansion of bone marrow-derived hMSCs on novel synthetic microcarriers in xeno-free, defined conditions. PLoS One 9(3):e92120

Hilkens P, Gervois P, Fanton Y, Vanormelingen J, Martens W, Struys T et al (2013) Effect of isolation methodology on stem cell properties and multilineage differentiation potential of human dental pulp stem cells. Cell Tissue Res 353(1):65–78

Hung SC, Chen NJ, Hsieh SL, Li H, Ma HL, Lo WH (2002) Isolation and characterization of size-sieved stem cells from human bone marrow. Stem Cells (Dayton, OH) 20(3):249–258

Iwase T, Nagaya N, Fujii T, Itoh T, Murakami S, Matsumoto T et al (2005) Comparison of angiogenic potency between mesenchymal stem cells and mononuclear cells in a rat model of hindlimb ischemia. Cardiovasc Res 66(3):543–551

Kang H-J, Kim H-S, Zhang S-Y, Park K-W, Cho H-J, Koo B-K et al (2004) Effects of intracoronary infusion of peripheral blood stem-cells mobilised with granulocyte-colony stimulating factor on left ventricular systolic function and restenosis after coronary stenting in myocardial infarction: the MAGIC cell randomised clinical trial. Lancet 363(9411):751–756

Kastrup J, Haack-Sorensen M, Juhl M, Harary Sondergaard R, Follin B, Drozd Lund L et al (2017) Cryopreserved off-the-shelf allogeneic adipose-derived stromal cells for therapy in patients with ischemic heart disease and heart failure-a safety study. Stem Cells Transl Med 6(11):1963–1971

Kawamoto A, Iwasaki H, Kusano K, Murayama T, Oyamada A, Silver M et al (2006) CD34-positive cells exhibit increased potency and safety for therapeutic neovascularization after myocardial infarction compared with total mononuclear cells. Circulation 114(20):2163–2169

Kim R, Lee S, Lee J, Kim M, Kim WJ, Lee HW et al (2018) Exosomes derived from MicroRNA-584 transfected mesenchymal stem cells: novel alternative therapeutic vehicles for cancer therapy. BMB Rep pii:4198

Kobayashi H, Ebisawa K, Kambe M, Kasai T, Suga H, Nakamura K et al (2018) Editors' choice effects of exosomes derived from the induced pluripotent stem cells on skin wound healing. Nagoya J Med Sci 80(2):141–153

Kondo M, Wagers AJ, Manz MG, Prohaska SS, Scherer DC, Beilhack GF et al (2003) Biology of hematopoietic stem cells and progenitors: implications for clinical application. Annu Rev Immunol 21(1):759–806

Kouroupis D, Churchman SM, McGonagle D, Jones EA (2014) The assessment of CD146-based cell sorting and telomere length analysis for establishing the identity of mesenchymal stem cells in human umbilical cord. F1000Res 3:126

Kumagai G, Tsoulfas P, Toh S, McNiece I, Bramlett HM, Dietrich WD (2013) Genetically modified mesenchymal stem cells (MSCs) promote axonal regeneration and prevent hypersensitivity after spinal cord injury. Exp Neurol 248:369–380

Kurtzberg J, Prockop S, Teira P, Bittencourt H, Lewis V, Chan KW et al (2014) Allogeneic human mesenchymal stem cell therapy (remestemcel-L, Prochymal) as a rescue agent for severe refractory acute graft-versus-host disease in pediatric patients. Biol Blood Marrow transplant 20(2):229–235

Kyodo (2018) https://www.japantimes.co.jp/news/2018/05/16/national/science-health/japan-oks-first-clinical-study-ips-cell-based-heart-treatment/#.W0NnWql9gdU

Lacerda JF, Martins C, Carmo JA, Lourenco F, Juncal C, Ismail S et al (2005) Haploidentical stem cell transplantation with purified CD34+ cells after a chemotherapy-alone conditioning regimen in heavily transfused severe aplastic anemia. Biol Blood Marrow transplant 11(5):399–400

Lan Y-W, Choo K-B, Chen C-M, Hung T-H, Chen Y-B, Hsieh C-H et al (2015) Hypoxia-preconditioned mesenchymal stem cells attenuate bleomycin-induced pulmonary fibrosis. Stem Cell Res Ther 6(1):97

Lauterboeck L, Saha D, Chatterjee A, Hofmann N, Glasmacher B (2016) Xeno-free cryopreservation of bone marrow-derived multipotent stromal cells from Callithrix jacchus. Biopreserv Biobank 14(6):530–538

Le PT-B, Van Pham P, Vu NB, Dang LT-T, Phan NK (2016) Expanded autologous adipose derived stem cell transplantation for type 2 diabetes mellitus. Biomed Res Ther 3(12):1034–1044

Lee OK, Kuo TK, Chen WM, Lee KD, Hsieh SL, Chen TH (2004) Isolation of multipotent mesenchymal stem cells from umbilical cord blood. Blood 103(5):1669–1675

Li WT, Leu YC, Wu JL (2010) Red-light light-emitting diode irradiation increases the proliferation and osteogenic differentiation of rat bone marrow mesenchymal stem cells. Photomed Laser Surg 28(Suppl 1):S157–S165

Li X, Shang Q, Zhang L (2014) Comparison of the efficacy of cord blood mononuclear cells (MNCs) and CD34+ cells for the treatment of neonatal mice with cerebral palsy. Cell Biochem Biophys 70(3):1539–1544

Li Y, Guo G, Li L, Chen F, Bao J, Shi YJ et al (2015) Three-dimensional spheroid culture of human umbilical cord mesenchymal stem cells promotes cell yield and stemness maintenance. Cell Tissue Res 360(2):297–307

Li J, Li H, Tian W (2018a) Isolation of murine adipose-derived stromal/stem cells using an explant culture method. Methods in molecular biology (Clifton, NJ) 1773:167–171

Li J, Curley JL, Floyd ZE, Wu X, Halvorsen YDC, Gimble JM (2018b) Isolation of human adipose-derived stem cells from lipoaspirates. Methods in molecular biology (Clifton, NJ) 1773:155–165

Liu Y, Chen X, Han W, Zhang Y (2017a) Tisagenlecleucel, an approved anti-CD19 chimeric antigen receptor T-cell therapy for the treatment of leukemia. Drugs of today (Barcelona, Spain: 1998) 53(11):597–608

Liu X, Li Q, Niu X, Hu B, Chen S, Song W et al (2017b) Exosomes secreted from human-induced pluripotent stem cell-derived mesenchymal stem cells prevent osteonecrosis of the femoral head by promoting angiogenesis. Int J Biol Sci 13(2):232–244

Liu T, Zhang Y, Shen Z, Zou X, Chen X, Chen L et al (2017c) Immunomodulatory effects of OX40Ig gene-modified adipose tissue-derived mesenchymal stem cells on rat kidney transplantation. Int J Mol Med 39(1):144–152

Locatelli F, Algeri M, Trevisan V, Bertaina A (2017) Remestemcel-L for the treatment of graft versus host disease. Expert Rev Clin Immunol 13(1):43–56

Lu D, Chen B, Liang Z, Deng W, Jiang Y, Li S et al (2011) Comparison of bone marrow mesenchymal stem cells with bone marrow-derived mononuclear cells for treatment of diabetic critical limb ischemia and foot ulcer: a double-blind, randomized, controlled trial. Diabetes Res Clin Pract 92(1):26–36

Mandai M, Kurimoto Y, Takahashi M (2017) Autologous induced stem-cell-derived retinal cells for macular degeneration. N Engl J Med 377(8):792–793

Mannon PJ (2011) Remestemcel-L: human mesenchymal stem cells as an emerging therapy for Crohn's disease. Expert Opin Biol Ther 11(9):1249–1256

Mareschi K, Biasin E, Piacibello W, Aglietta M, Madon E, Fagioli F (2001) Isolation of human mesenchymal stem cells: bone marrow versus umbilical cord blood. Haematologica 86(10):1099–1100

Massood E, Maryam K, Parvin S, Mojgan M, Noureddin NM (2013) Vitrification of human umbilical cord Wharton's jelly-derived mesenchymal stem cells. Cryo Letters 34(5):471–480

Mazo M, Gavira JJ, Abizanda G, Moreno C, Ecay M, Soriano M et al (2010) Transplantation of mesenchymal stem cells exerts a greater long-term effect than bone marrow mononuclear cells in a chronic myocardial infarction model in rat. Cell Transplant 19(3):313–328

Mehta A, Verma V, Nandihalli M, Ramachandra CJ, Sequiera GL, Sudibyo Y et al (2014) A systemic evaluation of cardiac differentiation from mRNA reprogrammed human induced pluripotent stem cells. PLoS One 9(7):e103485

Miyagi-Shiohira C, Kobayashi N, Saitoh I, Watanabe M, Noguchi Y, Matsushita M et al (2017) Evaluation of serum-free, xeno-free cryopreservation solutions for human adipose-derived mesenchymal stem cells. Cell Med 9(1–2):15–20

Miyamoto Y, Oishi K, Yukawa H, Noguchi H, Sasaki M, Iwata H et al (2012) Cryopreservation of human adipose tissue-derived stem/progenitor cells using the silk protein sericin. Cell Transplant 21(2–3):617–622

Mochizuki M, Nakahara T (2018) Establishment of xenogeneic serum-free culture methods for handling human dental pulp stem cells using clinically oriented in-vitro and in-vivo conditions. Stem Cell Res Ther 9(1):25

Monaco L, Faccio L (2017) Patient-driven search for rare disease therapies: the Fondazione Telethon success story and the strategy leading to Strimvelis. EMBO Mol Med 9(3):289–292

Negrin RS, Atkinson K, Leemhuis T, Hanania E, Juttner C, Tierney K et al (2000) Transplantation of highly purified CD34+Thy-1+ hematopoietic stem cells in patients with metastatic breast cancer. Biol Blood Marrow Transplant 6(3):262–271

O'Rourke C, Day AGE, Murray-Dunning C, Thanabalasundaram L, Cowan J, Stevanato L et al (2018) An allogeneic 'off the shelf' therapeutic strategy for peripheral nerve tissue engineering using clinical grade human neural stem cells. Sci Rep 8(1):2951

Oyekunle A, Koehl U, Schieder H, Ayuk F, Renges H, Fehse N et al (2006) CD34(+)-selected stem cell boost for delayed or insufficient engraftment after allogeneic stem cell transplantation. Cytotherapy 8(4):375–380

Panes J, Garcia-Olmo D, Van Assche G, Colombel JF, Reinisch W, Baumgart DC et al (2018) Long-term efficacy and safety of stem cell therapy (Cx601) for complex perianal fistulas in patients with Crohn's disease. Gastroenterology 154(5):1334–42.e4

Pellegrini G, Ardigo D, Milazzo G, Iotti G, Guatelli P, Pelosi D et al (2018) Navigating market authorization: the path holoclar took to become the first stem cell product approved in the European Union. Stem Cells Transl Med 7(1):146–154

Poltavtseva RA, Nikonova YA, Selezneva II, Yaroslavtseva AK, Stepanenko VN, Esipov RS et al (2014) Mesenchymal stem cells from human dental pulp: isolation, characteristics, and potencies of targeted differentiation. Bull Exp Biol Med 158(1):164–169

Prasad VK, Lucas KG, Kleiner GI, Talano JA, Jacobsohn D, Broadwater G et al (2011) Efficacy and safety of ex vivo cultured adult human mesenchymal stem cells (prochymal) in pediatric patients with severe refractory acute graft-versus-host disease in a compassionate use study. Biol Blood Marrow Transplant 17(4):534–541

Prindull G, Prindull B, Meulen N (1978) Haematopoietic stem cells (CFUc) in human cord blood. Acta Paediatr Scand 67(4):413–416

Prockop DJ, Sekiya I, Colter DC (2001) Isolation and characterization of rapidly self-renewing stem cells from cultures of human marrow stromal cells. Cytotherapy 3(5):393–396

Qian Q, Qian H, Zhang X, Zhu W, Yan Y, Ye S et al (2012) 5-Azacytidine induces cardiac differentiation of human umbilical cord-derived mesenchymal stem cells by activating extracellular regulated kinase. Stem Cells Dev 21(1):67–75

Reardon S, Cyranoski D (2014) Japan stem-cell trial stirs envy. Nature 513(7518):287–288

Reich-Slotky R, Colovai AI, Semidei-Pomales M, Patel N, Cairo M, Jhang J et al (2008) Determining post-thaw CD34+ cell dose of cryopreserved haematopoietic progenitor cells demonstrates high recovery and confirms their integrity. Vox Sang 94(4):351–357

Roberts I, Baila S, Rice RB, Janssens ME, Nguyen K, Moens N et al (2012) Scale-up of human embryonic stem cell culture using a hollow fibre bioreactor. Biotechnol Lett 34(12):2307–2315

Rodrigues JP, Paraguassu-Braga FH, Carvalho L, Abdelhay E, Bouzas LF, Porto LC (2008) Evaluation of trehalose and sucrose as cryoprotectants for hematopoietic stem cells of umbilical cord blood. Cryobiology 56(2):144–151

Rogulska O, Petrenko Y, Petrenko A (2017) DMSO-free cryopreservation of adipose-derived mesenchymal stromal cells: expansion medium affects post-thaw survival. Cytotechnology 69(2):265–276

Rojewski MT, Fekete N, Baila S, Nguyen K, Furst D, Antwiler D et al (2013) GMP-compliant isolation and expansion of bone marrow-derived MSCs in the closed, automated device quantum cell expansion system. Cell Transplant 22(11):1981–2000

Roy S, Arora S, Kumari P, Ta M (2014) A simple and serum-free protocol for cryopreservation of human umbilical cord as source of Wharton's jelly mesenchymal stem cells. Cryobiology 68(3):467–472

Schimmer J, Breazzano S (2016) Investor outlook: rising from the ashes; GSK's European approval of strimvelis for ADA-SCID. Human gene therapy. Clin Dev 27(2):57–61

Schmal O, Seifert J, Schaffer TE, Walter CB, Aicher WK, Klein G (2016) Hematopoietic stem and progenitor cell expansion in contact with mesenchymal stromal cells in a hanging drop model uncovers disadvantages of 3D culture. Stem Cells Int 2016:4148093

Sheridan WP, Fox R, Begley C, Maher D, McGrath K, Juttner C et al (1992) Effect of peripheral-blood progenitor cells mobilised by filgrastim (G-CSF) on platelet recovery after high-dose chemotherapy. Lancet 339(8794):640–644

Shivakumar SB, Bharti D, Subbarao RB, Jang SJ, Park JS, Ullah I et al (2016) DMSO- and serum-free cryopreservation of Wharton's jelly tissue isolated from human umbilical cord. J Cell Biochem 117(10):2397–2412

Silverman E (2018) Kymriah: a sign of more difficult decisions to come. Managed care (Langhorne, Pa) 27(5):17

Somlo G, Sniecinski I, Odom-Maryon T, Nowicki B, Chow W, Hamasaki V et al (1997) Effect of CD34+ selection and various schedules of stem cell reinfusion and granulocyte colony-stimulating factor priming on hematopoietic recovery after high-dose chemotherapy for breast cancer. Blood 89(5):1521–1528

Stirnadel-Farrant H, Kudari M, Garman N, Imrie J, Chopra B, Giannelli S et al (2018) Gene therapy in rare diseases: the benefits and challenges of developing a patient-centric registry for Strimvelis in ADA-SCID. Orphanet J Rare Dis 13(1):49

Takahashi K, Yamanaka S (2006) Induction of pluripotent stem cells from mouse embryonic and adult fibroblast cultures by defined factors. Cell 126(4):663–676

Thomas ED (2005) Bone marrow transplantation from the personal viewpoint. Int J Hematol 81(2):89–93

Vaes B, Van't Hof W, Deans R, Pinxteren J (2012) Application of MultiStem(R) allogeneic cells for immunomodulatory therapy: clinical progress and pre-clinical challenges in prophylaxis for graft versus host disease. Front Immunol 3:345

Van Pham P (2016a) Clinical application of stem cells: an update 2015. Biomed Res Ther 3(02):483–489

Van Pham P (2016b) Stem cell drugs: the next generation of pharmaceutical products. Biomed Res Ther 3(10):857–871

Van Pham P, Vu NB, Pham VM, Truong NH, Pham TL, Dang LT et al (2014a) Good manufacturing practice-compliant isolation and culture of human umbilical cord blood-derived mesenchymal stem cells. J Transl Med 12:56

Van Pham P, Vu NB, Phan NL-C, Le DM, Truong NC, Truong NH et al (2014b) Good manufacturing practice-compliant isolation and culture of human adipose derived stem cells. Biomed Res Ther 1(4):21

Van Pham P, Truong NC, Le PT, Tran TD, Vu NB, Bui KH et al (2016a) Isolation and proliferation of umbilical cord tissue derived mesenchymal stem cells for clinical applications. Cell Tissue Bank 17(2):289–302

Van Pham P, Vu NB, Phan NK (2016b) Umbilical cord-derived stem cells (MODULATISTTM) show strong immunomodulation capacity compared to adipose tissue-derived or bone marrow-derived mesenchymal stem cells. Biomed Res Ther 3(06):687–696

Wan Safwani WKZ, Makpol S, Sathapan S, Chua KH (2012) 5-Azacytidine is insufficient for cardiogenesis in human adipose-derived stem cells. J Neg Results Biomed 11:3

Wang HY, Lun ZR, Lu SS (2011) Cryopreservation of umbilical cord blood-derived mesenchymal stem cells without dimethyl sulfoxide. Cryo Letters 32(1):81–88

Wang C, Xiao R, Cao YL, Yin HY (2017) Evaluation of human platelet lysate and dimethyl sulfoxide as cryoprotectants for the cryopreservation of human adipose-derived stem cells. Biochem Biophys Res Commun 491(1):198–203

Wilson A, Trumpp A (2006) Bone-marrow haematopoietic-stem-cell niches. Nat Rev Immunol 6(2):93

Winter JM, Jacobson P, Bullough B, Christensen AP, Boyer M, Reems JA (2014) Long-term effects of cryopreservation on clinically prepared hematopoietic progenitor cell products. Cytotherapy 16(7):965–975

Xie CQ, Zhang J, Villacorta L, Cui T, Huang H, Chen YE (2007) A highly efficient method to differentiate smooth muscle cells from human embryonic stem cells. Arterioscler Thromb Vasc Biol 27(12):e311–e312

Xue X, Liu Y, Zhang J, Liu T, Yang Z, Wang H (2015) Bcl-xL genetic modification enhanced the therapeutic efficacy of mesenchymal stem cell transplantation in the treatment of heart infarction. Stem Cells Int 2015:176409

Yanagihara K, Uchida S, Ohba S, Kataoka K, Itaka K (2018) Treatment of bone defects by transplantation of genetically modified mesenchymal stem cell spheroids. Mol Ther Meth Clin Dev 9:358–366

Yuan Y, Yan G, Gong R, Zhang L, Liu T, Feng C et al (2017) Effects of blue light emitting diode irradiation on the proliferation, apoptosis and differentiation of bone marrow-derived mesenchymal stem cells. Cell Physiol Biochem 43(1):237–246

Zanata F, Bowles A, Frazier T, Curley JL, Bunnell BA, Wu X et al (2018) Effect of cryopreservation on human adipose tissue and isolated stromal vascular fraction cells: in vitro and in vivo analyses. Plastic Reconstruct Surg 141(2):232e–243e

Zhang W, Bai X, Zhao B, Li Y, Zhang Y, Li Z et al (2018) Cell-free therapy based on adipose tissue stem cell-derived exosomes promotes wound healing via the PI3K/Akt signaling pathway. Exp Cell Res pii:S0014-4827(18)30375-6

Zhou L, Song Q, Shen J, Xu L, Xu Z, Wu R et al (2017) Comparison of human adipose stromal vascular fraction and adipose-derived mesenchymal stem cells for the attenuation of acute renal ischemia/reperfusion injury. Sci Rep 7:44058

Chapter 7
Off-the-Shelf Mesenchymal Stem Cell Technology

Ngoc Bich Vu, Phuong Thi-Bich Le, Nhat Chau Truong, and Phuc Van Pham

Abbreviations

ADSC	Adipose derived stem cells
BM	Bone marrow
CHMP	The Committee for Medicinal Products for Human Use
GMP	Good manufacturing practice
HLA	Human leukocyte antigen
IMDM	Iscove's Modified Dulbecco's medium
MNC	Mononuclear cells
MSC	Mesenchymal stem cell
PRP	Platelet rich plasma
SVF	Stromal vascular fractions
UC	Umbilical cord

N. B. Vu
Laboratory of Stem Cell Research and Application, VNUHCM University of Science, Ho Chi Minh City, Vietnam

Stem Cell Institute, VNUHCM University of Science, Ho Chi Minh City, Vietnam

P. T.-B. Le
Van Hanh General Hospital, Ho Chi Minh City, Vietnam

N. C. Truong
Laboratory of Stem Cell Research and Application, VNUHCM University of Science, Ho Chi Minh City, Vietnam

P. Van Pham (✉)
Laboratory of Stem Cell Research and Application, VNUHCM University of Science, Ho Chi Minh City, Vietnam

Stem Cell Institute, VNUHCM University of Science, Ho Chi Minh City, Vietnam

Faculty of Biology-Biotechnology, VNUHCM University of Science, Ho Chi Minh City, Vietnam
e-mail: pvphuc@hcmuns.edu.vn; phucpham@sci.edu.vn

P. V. Pham (ed.), *Stem Cell Drugs - A New Generation of Biopharmaceuticals*, Stem Cells in Clinical Applications, https://doi.org/10.1007/978-3-319-99328-7_7

Mesenchymal Stem Cells

History and Characteristics

Mesenchymal stem cells (MSCs) are the most popular stem cells of the human body in terms of their distinct advantages for use in therapy. They can be isolated from a variety of tissues and can differentiate into a variety of cell types. The term "mesenchymal stem cells" was coined by Caplan in 1991 (Caplan 1991), though some discoveries about MSCs had already been demonstrated as early as the nineteenth century by Goujon (1869) {E., 1869 #13}, Tavassoli and Crosby (1968) (Schofield 1978). Friedenstein et al. performed a series of seminal studies in the 1960s and 1970 to demonstrate the osteogenic potential of a small population of cells in the bone marrow. They observed that these cells were so different from other hematopoietic cells due to their adherence to culture vessels and that they exhibited a morphology similar to fibroblasts (Friedenstein et al. 1970). In in vivo experiments, Friedenstein et al. showed that these cells could form skeletal tissues, including bone, cartilage, and adipose tissue. Therefore, they termed them "bone marrow stromal stem cells" (Owen 1988). To date, MSCs can be isolated from a variety of tissues in the human body, such as adipose tissue (Bernacki et al. 2008; van Vollenstee et al. 2016; Boquest et al. 2006), umbilical cord blood (Lee 2004; Bieback 2004; Pranke and Canabarro 2008), umbilical cord tissue (Falcon-Girard et al. 2013; Harris 2013; Van et al. 2016), Wharton's jelly (Ranjbaran et al. 2018; Ducret et al. 2016; Kargozar et al. 2018; Davies et al. 2017), and peripheral blood (Pieper et al. 2017).

Although MSCs from different tissues have different characteristics, they display common phenotypes and characteristics with only minor variations. As suggested by Dominici et al. (2006), and according to the International Society for Cellular Therapies (ISCT), the following are some minimal criteria of MSCs:

Markers

MSCs express CD105, CD73, and CD90, and are negative for CD45, CD34, CD14 or CD11b, CD79a or CD19, and HLA-DR. Besides these markers, Kolf et al. also suggested Stro-1 as a marker for MSCs (Kolf et al. 2007). Unfortunately, Stro-1 is not stable during culture and, therefore, should not be used as an MSC marker. Some other suggested MSC markers are stage-specific embryonic antigen 1 (SSEA-1), SSEA-4 (expressed in primitive MSCs) (Anjos-Afonso and Bonnet 2006; Gang et al. 2007), and CD106 or VCAM-1 (Carter and Wicks 2001). MSCs should be negative for the expression of CD11b and glycophorin-A (Pittenger 1999; Prockop et al. 2001).

Multilineage Differentiation Potential

MSCs are defined by their potential in vitro differentiation under suitable conditions into three kinds of cells of the mesoderm, including osteoblasts, chondrocytes, and adipocytes. Besides these three kinds of cells, MSCs can be transdifferentiated into some cell lines of endoderm or ectoderm, such as beta cells, neurons, and cardiomyocytes.

Shape and Adherence

Unlike hematopoietic cells, MSCs can adhere easily to plastic vessel surfaces and exhibit fibroblast-like morphology (Fig. 7.1).

Mesenchymal Stem Cells Are Suitable for Off-the-Shelf Products

Besides being a popular stem cell source, MSCs possess other characteristics that garner them as an ideal allogenic stem cell source. These characteristics include low immunogenicity and immune modulation capability. Indeed, MSCs express very low levels of MHC class I, and do not express MHC class II. More importantly, they cannot activate allogenic lymphocytes (Koç and Gerson 2003; Berglund et al. 2017).

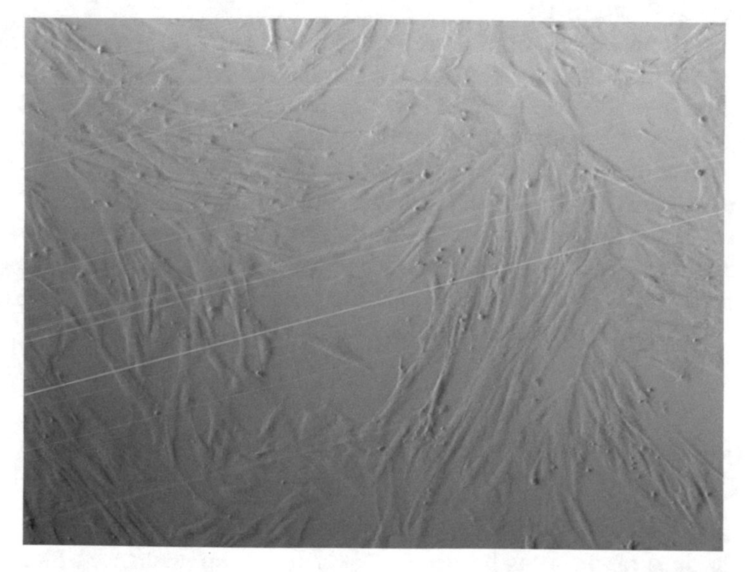

Fig. 7.1 The umbilical cord derived mesenchymal stem cells displayed the fibroblast like cells in the flast surface (×100)

Moreover, MSCs can also inhibit the proliferation of syngeneic and allogeneic T lymphocytes in a manner independent of MHC expression (Nauta 2006).

The immune modulation capacity of MSCs has been observed in numerous studies (Li et al. 2010; Bifari et al. 2008; Knaän-Shanzer 2014; FIBBE et al. 2007; Hong et al. 2012; Ansboro et al. 2017; Semedo et al. 2009). There are two ways that MSCs can modulate the host immune system; the first is immunomodulation by soluble factors and the second is immunomodulation by cell-cell contact. MSCs can produce a variety of soluble factors that can suppress immune cell proliferation in the host. Such factors include transforming growth factor-β1 (TGF-β1), prostaglandin E2 (PGE2), hepatocyte growth factor (HGF), indoleamine-pyrrole 2,3-dioxygenase (IDO), nitric oxide (NO), and interleukin (IL)-10). TGF-β1 is growth factor that exerts strong suppressor activity on immune cell proliferation (Du et al. 2018; Yoo et al. 2013). PGE2 can be upregulated in MSCs when cocultured with mononuclear cells; moreover, PGE2 can inhibit T cell proliferation (Jarvinen et al. 2008). IDO, also produced by MSCs, inhibits the growth and function of immune cells (Jarvinen et al. 2008). Some studies have reported that in an inflammatory microenvironment, MSCs can produce certain enzymes, such as cyclooxygenase 2 (COX-2), PGE2, and IDO, which also act as immune suppressors (Krampera et al. 2006; Ryan et al. 2007; DelaRosa et al. 2009).

In the other method (cell–cell contact between MSCs and immune cells), MSCs can inhibit immune cell proliferation. Some recent studies have shown that MSCs can inhibit immune cell proliferation by MSC–T cell contact (Han et al. 2011; Krampera 2002; English et al. 2009). In a study by Han et al. (2011), the authors showed that both embryonic stem cells (ESCs) and MSCs exert immunosuppressive effects. However, the activity did not depend on the concentration of certain cytokines (e.g., TGF-β or IDO).

It is worth noting that MSCs can be suppressed by T-cell production of IL-2, IL-12, interferon gamma (IFN-γ), tumor necrosis factor alpha (TNF-α), IL-4, IL-5, IL-1β, and IL-10. Han et al. showed that the population of Foxp3(+) regulatory T cells significantly increased when MSCs or ESCs contacted with T cells. From these observations, Han et al. suggested that MSCs can contact with T cells and cause an increase of Foxp3(+) regulatory T cells that can inhibit immune responses (Han et al. 2011). Similarly, Krampera (2002) studied immune suppression of MSCs in directly and indirectly cocultured MSCs using a Transwell system with cocultured T cells. The results confirmed that the inhibitory effect of MSCs was reduced when MSCs were indirectly cocultured with T cells (Krampera 2002).

Therapeutic Mechanisms of Off-the-Shelf Mesenchymal Stem Cells

There are three mechanisms by which MSCs are useful as therapy. Firstly, after transplantation MSCs can home to sites of injured tissues and once there, they can be differentiated into tissue specialized cells following stimulation by certain endogenous factors. Secondly, MSCs can produce a range of growth factors that stimulate neoangiogenesis, promote self-renewal of endogenous stem cells, inhibit fibrosis, and inhibit apoptosis. Thirdly, MSCs can efficiently suppress the local inflammatory via their secretome as well as direct cell–cell contact.

Indeed, off-the-shelf MSC products exert their therapeutic effects through the action of their secretomes. In addition to facilitating angiogenesis and self-renewal, while inhibiting fibrosis and apoptosis, secretomes can suppress local inflammatory responses via suppression of lymphocytes. Although MSCs exhibit low immunogenicity, they are still rejected from the host after about 6 months. During the 6 months in the body, MSCs exert their therapeutic effects by communication with local cells and other stem cells (via cytokines which they secrete or via direct contact with pertinent target cells).

However, perhaps the main reason why MSCs have been approved as off-the-shelf mesenchymal stem cell products is their immune modulation capacity (Table 7.1). In 2012, the first off-the-shelf mesenchymal stem cells were approved in Canada to treat graft-versus-host disease. This product (called "Prochymal") contained MSCs derived from allogenic human bone marrow and showed long-term proliferation. This product relied on the immune modulation capacity of MSCs, such as their ability to suppress GVHD responses (Prasad et al. 2011; Martin et al. 2010; Kurtzberg et al. 2010, 2014; Gennery 2016).

Similarly, the product "Temcell HS" (developed in Japan) contains MSCs from bone marrow and was approved in Japan for GVHD treatment in 2016 (Najima and Ohashi 2017). In 2014, another off-the-shelf mesenchymal stem cell product (under the trade name "Cartilatist") was approved, in Korea this time, to treat knee osteoarthritis (Park et al. 2017). Recently, the first off-the-shelf mesenchymal stem cell product (Cx601, darvadstrocel) received a positive CHMP opinion to treat complex perianal fistulas in Crohn's disease in Europe (Sheridan 2018). Thus, all these above products act as immune suppressors in the host, suggesting that the main mechanism of off-the-shelf MSCs is likely immune modulation.

Table 7.1 Some off-the-shelf mesenchymal stem cell products approved in various countries

Names of products	Component of stem cells	Indications	Company	Country
Cartistem	MSCs from UCB	OA	Medipost	Korea
Prochymal	MSCs from BM	GVHD	Osiris Therapeutics	Canada
AlloStem	MSCs from BM	OA	AlloSource	USA
Osteocel Plus	MSCs from BM	OA	NuVasive	USA
Trinity Evolution	MSCs from BM	OA	Orthofix	USA
Cx601 (Darvadstrocel)	MSCs from AT	Complex perianal fistulas in Crohn's disease	Takeda and TiGenix	Europe (nearly approved)

AT adipose tissue, *BM* bone marrow, *MSCs* mesenchymal stem cells, *UCB* umbilical cord blood, *Auto* autologous, *allo* allogenic, *OA* osteoarthritis

Although recently there have been some approved off-the-shelf MSCs for treatment of injuries unrelated to the immune system, few products have been approved for immune system indications (or the products have been largely unproven). Prochymal was used to treat myocardial infarction (MI), in 2010, in a randomized, double-blind, placebo-controlled clinical trial (Gersh 2010). In this study, Prochymal was intravenously transfused into 53 MI patients at different doses: 0.5, 1.6, and five million cells/kg. The results showed that the global symptom score in all the patients and the ejection fraction in the important subset of anterior MI patients were both significantly better ($p = 0.027$) after hMSC-treatment (i.e., Prochymal treatment) as compared with placebo treatment (Gersh 2010).

In 2017, the first-in-human clinical trial using off-the-shelf MSCs from adipose tissue to treat intramyocardial injection in ten patients was conducted in Denmark (Kastrup et al. 2017). In fact, off-the-shelf MSC products for this clinical trial were produced from adipose tissue in bioreactors without the use of animal constituents. They were cryopreserved and stored in vials in nitrogen dry-storage containers until use. All participants were injected with the MSCs in the myocardium. The results showed that four out of ten patients developed donor-specific de novo HLA class I antibodies, and two out of ten had donor-specific HLA antibodies already at baseline. However, there were no clinical symptoms in inflammatory parameters in the follow-up period. Moreover, after 6 months of treatment, left ventricular end systolic volume decreased and left ventricular ejection fraction increased. Moreover, these changes were independent of the presence or absence of HLA antibodies (Kastrup et al. 2017).

Besides some off-the-shelf MSCs derived from bone marrow and adipose tissue, there is a product with the trade name "Modulatist" that was produced from umbilical cord tissue and developed in Vietnam (Van Pham et al. 2016). Modulatist showed strong immunomodulation capacity compared to adipose tissue-derived or bone marrow-derived MSCs (Van Pham et al. 2016). This product was clinically used in a case report for chronic obstructive pulmonary disease (COPD) (Le et al. 2016). In this study, two patients were intravenously infused with 10^6 cells/kg and then evaluated by the COPD assessment test (CAT) score as well as the Modified Medical Research Council Dyspnea Scale (mMRC) score after transplantation (1, 3, and 5 months post-transplantation). The results showed that Modulatist significantly improved severe COPD, especially after 3 months (Le et al. 2016).

Off-the-Shelf Mesenchymal Stem Cell Technology

Isolation of Mesenchymal Stem Cells

Isolation of MSCs is the first step in off-the-shelf MSC technology. Although MSCs can be derived from various tissues, off-the-shelf stem cells are usually isolated from four tissues, namely bone marrow, adipose tissue, umbilical cord blood and umbilical cord tissue. The MSCs from these tissues are used to successfully produce the off-the-shelf MSCs. However, these sources have different advantages and disadvantages.

Bone marrow-derived MSCs were likely the first discovered source MSCs and, therefore, have an established timeline and milestones for their applications in the clinic. Notably, bone marrow-derived MSCs are the first kind of MSCs to be approved as off-the-shelf MSCs in Canada in 2012, under the trade name "Prochymal," for GVHD treatment.

Umbilical cord blood-derived MSCs have also been successfully used to produce a stem cell drug, under the name "Cartistem"; it was approved in Korea for knee osteoarthritis in January 2012. Currently, adipose-derived stem cells (ADSCs) are used to develop "Alofisel," a product produced by Tigenix and Takeda. Alofisel is the first product to be approved in Europe to treat complex perianal fistulas in Crohn's disease. Umbilical cord-derived stem cells are novel sources of MSCs; the first application using this source as off-the-shelf technology was for the treatment of COPD in Vietnam; the product was named Modulatist (Le et al. 2016).

Isolation of MSCs from BM

MSCs were first isolated in the 1970s by Fridenstein et al. (1974). To date, the procedure of BM-MSC isolation has continuously improved and now MSCs can be expanded to greater numbers and to clinical standards. There are two main steps of MSC isolation and expansion from BM.

In the first step, the mononuclear cells (MNCs) are enriched or isolated from whole bone marrow fluid by gradient centrifugation with Ficoll 1.077. With some procedures, bone marrow blood can be diluted with PBS before it is used to isolate MNCs. Enrichment of MNCs by red blood cell lysis was introduced as the most efficient method in the MSC isolation process (Horn et al. 2008, 2011). In another protocol, MNCs can automatically be enriched using equipment such as Biosafe Sepax (O'Connor et al. 2007).

In the second step, MNCs are cultured to select for adherent cells which will become enriched for MSCs. The popular medium used in this procedure is Dulbecco's Modified Eagle Medium: Nutrient Mixture F-12 (DMEM/F12) (1:1), although some procedures have also used DMEM or alpha Minimum Essential Medium (MEM) (Pytlík et al. 2009). Unlike traditional culture medium that is supplemented with 10–20% fetal calf serum (FCS) or fetal bovine serum (FBS), in good manufacturing practices (GMP)-compliant procedures, the medium is supplemented with xeno-free components or some defined factors.

In the xeno-free media approach, autologous or allogeneic plasma (or serum) is used. Plasma or serum is usually collected from peripheral blood, but in some cases, they can be obtained from umbilical cord blood (Esmaeli et al. 2016), which also efficiently stimulates in vitro MSC proliferation (Blázquez-Prunera et al. 2017). For serum or plasma, the serum or plasma from AB blood donor is used

(Kocaoemer et al. 2007). Recently, the platelet lysate (PL) (Lange et al. 2007; Bernardo et al. 2007) or platelet-rich plasma (PRP) (Van Pham et al. 2014a, b, c; Kocaoemer et al. 2007) has replaced plasma or serum, in order to reduce the proteins in plasma or serum that can affect the MSCs or further applications. Plasma, serum, PL, or PRP can be added to the completed medium at 2–10% concentration (Fekete et al. 2012).

Almost all studies have confirmed that BM-MSCs cultured in medium with plasma, serum, PL, or PRP exhibit the phenotypes similar to BM-MSCs cultured in traditional conditions with FBS or FCS. Indeed, they also maintained MSC characteristics, such as adherence to plastic vessels with fibroblast-like shape, expression of particular makers of MSCs (CD44, CD73, CD90), and absence of expression of certain hematopoietic markers (CD14, CD34, CD45, HLA-DR). They can also be differentiated into trilineage of mesoderm cells, such as osteoblasts, chondrocytes and adipocytes (Lange et al. 2007; Pérez-Ilzarbe et al. 2009). In the xeno-free culture conditions, BM-MSCs also maintained their immune modulating capability, including inhibition of T cells and production of TNF-alpha and IFN-gamma (Bernardo et al. 2007). Importantly, spontaneous cell transformation was not observed in xeno-free medium culture (Bieback et al. 2009).

Some commercially available defined serum-free media have also been developed in recent years. Gottipamula et al. (2014) used BD Mosaic™ Mesenchymal Stem Cell Serum-Free media (BD-SFM) (BD Biosciences, San Jose, CA, USA) and Mesencult-XF (MSX) (Stemcell Technologies, Köln, Germany) to isolate and grow BM-MSCs (Gottipamula et al. 2014). The results showed that both these media could support BM-MSC growth (Gottipamula et al. 2014). In another study, Gottipamula et al. (2013) compared five different kinds of serum-free media for BM-MSCs, including StemPro MSC SFM Xeno-free TM (Gibco/Invitrogen, Karlsruhe, Germany), StemPro MSC SFM TM (Gibco/Invitrogen), Mesencult-XFTM (Stemcell Technologies, Canada), BD Mosaic TM Mesenchymal Stem Cell Serum-Free media (BD-SFM), and TheraPEAKTM MSCGM-CD TM (Lonza) (Gottipamula et al. 2013). The authors showed that BD Mosaic™ Mesenchymal Stem Cell Serum-free (BD-SFM) medium is suitable to use in large-scale cultures of BM-MSCs (Table 7.2).

Isolation of MSCs from Umbilical Cord Blood

Umbilical cord blood is a source rich of various kinds of stem cells, including MSCs, hematopoietic stem cells, and endothelial progenitor cells. MSCs from umbilical cord blood (UCB-MSCs) was first reported by Lee et al. (2004). In this study, Lee et al. showed that UCB-MSCs display a phenotype similar to BM-MSCs. UCB-MSCs could also be differentiated into osteoblasts, chondrocytes, and adipocytes (Lee 2004). Moreover, UCB-MSCs could be isolated from cryopreserved umbilical cord blood samples (Lee et al. 2004; Fujii et al. 2017).

Similar to BM-MSCs, UCB-MSCs are isolated by a 2-step protocol. In the first step, MNCs are isolated or enriched, and in the second step, UCB-MSCs are cultured.

Table 7.2 Some commercialized products for mesenchymal stem cell culture and expansion

No.	Products/technologies	Companies
1	BD Mosaic™ Mesenchymal Stem Cell Serum-Free media (BD-SFM)	BD Bioscience, USA
2	Mesencult-XF (MSX)	Stemcell Technologies, Canada
3	StemPro MSC SFM XenoFree	Invitrogen, USA
4	StemPro MSC SFM	Invitrogen, USA
5	TheraPEAKTM MSCGM-CD TM	Lonza, USA
6	MSCCult	Regenmed Lab., VN
7	ADSCCult	Regenmed Lab., VN
8	MSCCult Pro	Regenmed Lab., VN
9	StemGold MSC XF Medium	Atlantis Bioscience Pte Ltd., Singapore
10	Mesenchymal Stem Cell (MSC) Medium Kit	Atlantis Bioscience Pte Ltd., Singapore
11	KBM ADSC-1	Atlantis Bioscience Pte Ltd., Singapore
12	MSC NutriStem® XF Medium	Biological Industries USA

In the first procedure, UCB is diluted with PBS and then centrifuged in the gradient Ficoll 1.077 to obtain the interphase that contains the MNCs. Alternatively, for the interphase collection, some automatic systems can be used, such as Sepax (Biosafe/GE Healthcare, Eysins, Switzerland), AutoXpress® Platform ("AXP®") and BioArchive System (Cesca Therapeutics, Rancho Cordova, California, USA), PrepaCyte®-CB (Cryo-Cell International, Oldsmar, Florida, USA), and Cord Blood 2.0 TM (Americord, New York, NY, USA), to enrich for MNCs.

In the next procedure, MNCs are cultured on a vessel surface to select for adherent cells. The popular basal medium used to culture these cells is Iscove's Modified Dulbecco's Medium (IMDM) (Lee 2004; Divya et al. 2012). Some other studies have also used alpha-MEM (Hildebrandt et al. 2009; Kim et al. 2015; Jung et al. 2015) or DMEM/F12 (Shetty et al. 2007) to culture these cells. For use of these cells in the clinic, the media should be supplemented with some xeno-free components or defined factors, similar to culture of BM-MSCs.

Liu et al. (2007) developed a serum-free medium for UCB-MSCs based on IMDM basal medium supplemented with fibroblast growth factor (FGF) (17.91 ng/mL), human albumin (2.80 mg/mL), and hydrocortisone (27.65 μM) (Liu et al. 2007). Cells cultured in this condition retained their differentiation potential, that is, the ability to differentiate in vitro into mesenchymal lineages, including chondrocytes, adipocytes, and osteoblasts (Liu et al. 2007). Using another approach, Van Pham et al. (2014a, b, c) replaced FBS with PRP (obtained from the same umbilical cord blood samples) to culture UCB-MSCs (Van Pham et al. 2014a, b, c). The results confirmed that UCB-MSCs could be obtained by culturing MNCs in IMDM medium supplemented with PRP from the same blood samples (Van Pham et al. 2014a, b, c; Van Pham and Phan 2014).

Isolation of MSCs from Umbilical Cord Tissues

UC-MSCs can be isolated from UC tissues by fragment expansion procedure or single cell culture procedure. In both procedures, UC tissues are cut into small fragments about 1–2 mm^2 in size. In the first procedure, tissue fragments are seeded onto vessel surfaces (Van Pham et al. 2015). After 7–14 days of incubation, the UC-MSC candidates will migrate out from the pieces. The cells are then subcultured to be expanded in the next step. For the second procedure, small tissue fragments are minced (Hassan et al. 2017) or digested by enzymes, such as collagenase, to obtain single cells (Han et al. 2013; Beeravolu et al. 2017). These single cells are then collected and expanded in culture vessels. Regarding culture medium, UC-MSCs can be isolated and cultured in low glucose (LG)-DMEM supplemented with 2%, 5% or 10% PL (Smith et al. 2016). Smith et al. (2016) showed that 10% PL in the medium was optimal for UC-MSCs. The cells in this medium were smallest and most viable, expressed the typical markers of MSCs, showed high colony forming efficiency, and exhibited trilineage differentiation (Smith et al. 2016).

UC-MSCs can also be cultured in medium supplemented with PRP. In DMEM/F12 supplemented with 2.5%, 5%, 7.5% or 10% PRP, UC-MSCs showed good proliferation; however, with 7.5% or 10% PRP, growth of UC-MSCs was significantly greater compared to that with 2.5% or 5% PRP. At the higher concentrations of PRP, UC-MSCs maintained genomic stability with normal karyotype after 15 subcultures, maintained their differentiation potential, and failed to cause tumors in NOD/SCID mice (Van Pham et al. 2015).

UC-MSC have also been successfully isolated and cultured in alpha-MEM with 10% human serum (Hatlapatka et al. 2011). Under this condition, UC-MSCs displayed the in vitro immunoprivileged and immunomodulatory properties (Hatlapatka et al. 2011). Hartmann et al. (2010) evaluated the culture of UC-MSCs in serum-free media conditions using defined serum-free media (Hartmann et al. 2010). They used the StemPro MSC SFM medium, supplemented with 2% GMP-compliant human serum (Centre for Clinical Transfusion Medicine Tübingen gGmbH, Tübingen, Germany), and MesenCult ACF medium (Stemcell Technologies, Canada). UC-MSCs cultured in these conditions not only retained the MSCs characteristics but also exhibited the ability to suppress T cell proliferation; their suppression was stronger than that of UC-MSCs cultured in medium containing FBS (Hartmann et al. 2010).

In another strategy, Wu et al. (2016) were the first to expand UC-MSCs in a complete serum-free, xeno-free, chemically defined, and non-coated plate based culture system. In their work, Wu et al. (2016) used IMDM medium as the basal medium, and supplemented it with some factors, including human albumin serum.

Isolation of MSCs from Adipose Tissues

Adipose tissue is a rich resource of MSCs. As there are many advantages of using ADSCs, they have increasingly been studied for use in the clinic. Similar to UC-MSCs, ADSCs are isolated from solid tissues. Therefore, they can be isolated, too, by a short two-step procedure. In the first step, adipose tissues are digested by various methods to collect the stromal vascular fractions (SVFs). The SVFs are then cultured to isolate and enrich for the ADSCs in the second step. There are different ways to collect SVFs; one is an enzymatic method. In the enzymatic method, adipose tissues can be digested by collagenase (Zimmerlin et al. 2009; Francis et al. 2010) or lecithin (Van Pham et al. 2013; Tzouvelekis et al. 2011) to release the SVFs. In the next step, SVF cells are cultured in the basal medium supplemented with plasma, serum, PRP or defined factors. SVF cells can be cultured in DMEM/F12 supplemented with 10% PRP from peripheral blood (Van Pham et al. 2014a, b, c). Escobar and Chaparro (2016) was able to culture SVF cells in medium with 5% PL (Escobar and Chaparro 2016), or 10% human serum (Paula et al. 2015). It is worth noting that ADSCs in human serum overexpress the c-myc protein but bypass spontaneous cell transformation (Paula et al. 2015). Moreover, ADSCs can be successfully isolated and grown in a defined chemical medium (Lee et al. 2017; Rajala et al. 2010; Lindroos et al. 2009).

MSC Expansion and Proliferation

MSCs require adherence on plastic vessel surfaces in order to grow. Therefore, they need this surface for expansion. To date, there are three ways to scale up the MSC expansion phase to produce greater quantities of off-the-shelf MSC products (Table 7.3).

Scale-Up in T-Flasks

The first strategy relates to the use of big T-flasks (T-175 or T-225 cm^2) for culture and expansion. This is the easiest way to expand MSCs; however, the efficacy of scale-up is low. The number of MSCs increased, though nonsignificantly, in this method compared to other methods. To increase this efficacy, some multilayer flasks have been developed which increase the area for MSC adherence and proliferation.

Scale-Up Culture in Closed Hollow Fiber-Based Bioreactor (Quantum Cell Expansion System)

BM-MSCs can be successfully scaled up in the Quantum cell expansion system (Terumo, Japan). Rojewski et al. (2013) used the Quantum cell expansion system to expand BM-MSCs in media with FCS or PL. The results showed that both FSC and PL-supplemented media could expand BM-MSCs in the Quantum system. The authors succeeded to obtain an average of more than 100 × 10^6 MSCs from as little as 18.8–28.6 ml of BM aspirate using PL-supplemented media (Rojewski et al. 2013). BM-MSCs from this procedure exhibited the characteristic MSC phenotype,

Table 7.3 Some scale-up systems for MSC expansion

Flatform	Technology	Volume	Example technologies
Flask	T-flasks (single layer/≤5 multilayers)	30 mL/200 mL	Vented flasks (Corning)/Millicell® HY multilayer culture flasks (EMD Millipore)
	Multilayer stacks (10/40 stack layers)	1.4 L/5.5 L	CellSTACK® cell culture chambers (Corning), Nunc Cell Factory™ systems (Thermo Scientific)
	Closed system, multilayer stacks (12/36/120 layers)	1.3 L/3.9 L/13 L	HYPERFlask® Cell Culture Vessels (Corning)
	Pilot scale, static	19.8 L	Integrity™ Xpansion™ Multiplate (Pall Corporation)
Microcarrier-based culture system	Spinner flasks	125 mL to 3 L	Corning® Disposable Spinner Flasks (Corning)
	Mini-reactor systems	3–250 mL	DASbox®Mini Bioreactor System (Eppendorf), BioLevitator™ 3D Cell Culture System (Hamilton), TAP ambr™ microbioreactor (Sartorius), Micro-24 MicroReactor System (Pall Corporation)
	Benchtop stirred reactors	1–5 L	Mobius® CellReady (EMD Millipore), CelliGen® BLU (Eppendorf), UniVessel® SU (Sartorius)
	Pilot scale, stirred reactors	50–300 L	Mobius® CellReady (EMD Millipore), CelliGen® BLU (Eppendorf), BIOSTAT® STR (Sartorius), Xcellerex™ XDR (GE Healthcare), HyPerforma™ Single-use Bioreactor (Thermo Scientific), Nucleo™ Single-use Bioreactor (Pall Corporation), Allegro™ STR 200 (Pall Corporation)
	Production scale, stirred reactors	500–2000 L	BIOSTAT® STR (Sartorius), Xcellerex™ XDR (GE Healthcare), HyPerforma™ Single-use Bioreactor (Thermo Scientific), Nucleo™ Single-use Bioreactor (Pall Corporation)
	Oscillating motion reactors, surface aeration only	300–500 L	WAVE bioreactor™ system (GE Healthcare), BIOSTAT® RM (Sartorius), SmartBag™ containers (Finesse), Appliflex™ systems (Applicon), CELL-tainer® (CELLution Biotech/Lonza), XRS-20 Bioreactor System (Pall Life Sciences)
	Oscillating motion reactors, sparging	30–1000 L	BaySHAKE® (Bayer)
	Vertical wheel/bubble column	50–500 L	Vertical-Wheel™ reactor (PBS), CellMaker PLUS™ system (Cellexus)

(continued)

Table 7.3 (continued)

Flatform	Technology	Volume	Example technologies
Hollow-fiber-based bioreactor	Hollow fiber	4 L	Quantum Cell Expansion System (Terumo BCT)

fulfilled the minimal criteria of MSCs (Rojewski et al. 2013), maintained their T lymphocyte-inhibitory capacity (Nold et al. 2013), and retained their genetic stability (Jones et al. 2013).

This system was also used to produce ADSCs under GMP compliant conditions. The authors compared the culture efficacy of ADSCs in media supplemented with 5% PL versus 10% FBS. The results showed that after 30×10^6 ADSCs were loaded into the bioreactor, after 17 days the yield of ASCs was 546×10^6 ASCs in PL-supplemented medium, as compared to 111×10^6 ASCs in FBS-supplemented medium. They also showed that ADSCs fulfilled the ISCT criteria for MSCs, and demonstrated genomic stability and sterility (Haack-Sørensen et al. 2018). In an earlier study, Haack-Sørensen et al. (2018) also showed that expansion of ADSCs in the Quantum cell expansion system significantly yielded greater cell numbers than those grown in T-flasks; the ADSCs were found to be pure and safe for clinical applications (Haack-Sørensen et al. 2016).

Scale-Up in Microcarrier-Based Culture System

The microcarrier-based culture system is a culture system whereby MSCs adhere to a carrier (microbeads). The microbeads are then suspended/floated in the medium. By this method, the area of MSC adherence is significantly increased. Beads cultured with MSCs can be incubated in medium in a spinner flask bioreactor (dos Santos et al. 2011), wave bioreactor, stirred-tank bioreactor (dos Santos et al. 2014), or magnetic bioreactor. By this method, dos Santos et al. (2011) could expand BM-MSCs by 18 ± 1-fold and ADSCs by 14 ± 7-fold compared the traditional culture. These cells also maintained the minimal criteria of MSCs (dos Santos et al. 2011).

Using the stirred-tank bioreactor, dos Santos et al. (2014) also successfully expanded MSCs from BM and adipose tissue. In xeno-free conditions, the yield for BM-MSCs and ASCs, after 7 days, was $(1.1 \pm 0.1) \times 10^8$ and $(4.5 \pm 0.2) \times 10^7$ cells, respectively. More importantly, these MSCs retained the MSC phenotypes (such as positive expression of CD73, CD90, and CD105, and negative expression for CD31, CD80, and HLA-DR), and multilineage differentiation potential toward osteoblasts, chondrocytes, and adipocytes (dos Santos et al. 2014). Similarly, Carmelo et al. (2015) succeeded to expand BM-MSCs and ADSCs in a similar system (stirred-tank bioreactor). They achieved maximal cell densities of 3.6×10^5 and 1.9×10^5 cells/mL for BM-MSCs (0.60 ± 0.04 per day) and ASCs (0.9 ± 0.1 per day) cultures, respectively, following 7 and 8 days of culture, respectively (Carmelo et al. 2015).

The stirred-tank bioreactor in spinner flask culture system also can be used to expand UC-MSCs in xeno-free culture (Mizukami et al. 2016). This system permitted production of 2.4 (±1.1) × 10^5 cells/mL ($n = 4$) after 5 days of culture, corresponding to a 5.3 (±1.6)-fold increase in cell number. In the stirred tank bioreactor (800 mL), MSC-UCs could reach a yield of 115 million cells after 4 days. UC-MSCs still retained their phenotypes, differentiation potential, and immune modulation capacity (Mizukami et al. 2016). Recently, Lawson et al. (2017) scaled up BM-MSC expansion in a 50-L bioreactor for 11 days using medium alpha-MEM supplemented with 5% PL. Lawson et al. achieved a 43-fold expansion of MSCs with a final yield of 1.28 × 10^{10} cells. These cells displayed similar properties as those grown in flasks (Lawson et al. 2017).

Control of MSC Quality and Safety

The control MSC quality during the expansion is the essential step to make sure the stable quality between batch to batch. Depending on the some different technologies, some standards also vary from technology to technology. However, some common minimal criteria of MSCs are listed below.

MSC Quality Control

The first aspect, good quality, relates to MSC characteristics. After a long term of of expansion, MSCs should maintain their characteristic phenotype, without any spontaneous differentiation. The quality of MSCs is controlled by the properties listed below:

Self-renewal: Self-renewal is evaluated by a clonogenicity assay. In this assay, MSCs are cultured in a petri dish at low density to permit MSCs to form colonies. This is a simple, inexpensive, and highly reproducible assay.

Differentiation: Differentiation potential is an important assay to determine the stemness of MSCs. This also is a criterion of MSCs that was first suggested by Dominici et al. (2006). MSCs must be able to differentiate into three kinds of mesodermal cells, namely osteoblasts, adipocytes, and chondroblasts.

Maintenance of MSC phenotype: After long-term expansion, MSCs should maintain the characteristic MSC phenotype, as suggested by Dominici et al. (2006), which includes positive markers (CD13, CD44, CD73, CD90, and CD105) and negative markers (CD14, CD34, CD45, and HLA-DR).

Low or nonexpression of senescent phenotype: Senescence is the most important issue affecting MSCs in off-the-shelf MSC production. Indeed, MSCs can proceed to aging after 20–50 doublings, depending on the culture medium as well as cell source. The senescent phenotypes are characterized by various assays, such as aneuploidy (which can be evaluated by karyotyping) and accumulation of β-galactosidase (which can be detected by β-galactosidase staining assay). Senescence of MSCs can also be detected by mutations in certain genes (e.g., p53), upregulation of certain genes (hyaluronan, proteoglycan link protein 1, keratin 18, brain-derived neurotrophic factor, renal tumor antigen, etc.), downregulation of genes (e.g., pleiotrophin), and reduction of differentiation potential.

MSC Safety

In vitro expansion of MSCs may encounter safety issues such as (1) contamination by viruses, including hepatitis B and C, HIV, human T-cell leukemia virus type 1, and syphilis; (2) contamiation by bacteria, fungi, and mycoplasma; (3) contamination by endotoxins; and (4) tumorigenicity of expanded MSCs.

To reduce these issues, the MSC expansion procedure should follow the GMP guidelines with some control points during the MSC production from tissue screening to MSC cryopreservation.

- Tissues used to isolate stem cells should be carefully checked for contamination of hepatitis B and C, HIV, human T-cell leukemia virus type 1, and syphilis, both in the case of tissue samples and donors. At present, all tissues and donors should be screened in compliance with the blood bank guidelines. Only samples of both tissues and donors that test negative for all viruses are used for further processing.
- During the process, and at the final stage of stem cell collection, MSCs and the conditioned media should be collected and checked for bacteria and fungi contamination. The bacterial and fungal contamination of classical pharmaceutical products is excluded by standardized tests, as set in Europe (Pharmacopoeia [EP] 2005) or in the USA (Pharmacopoeia [USP] 2011). Mycoplasma contamination also should be checked according to European, US, and Japanese pharmacopeia guidelines.
- Endotoxins are lipopolysaccharides from gram-negative bacteria that often cause some serious health problems, such as diarrhea, septic shock, and marrow necrosis. Endotoxin testing seems to be the essential test for cellular products before these products would be used in human beings. The endotoxin limit for cellular products is usually 3.0 EU/kg/dose. Endotoxin testing is generally carried out by the Limulus amebocyte lysate method; it is also carried out by some commercialized kits.
- Tumorigenicity is the most concerning safety issue when using expanded MSCs for clinical applications, although to date there have been no reports about this serious effect in expanded MSCs. Nowadays, the tumorigenicity of MSCs can be checked by transplantation of a number of MSCs to NOD/SCID mice for several months with close monitoring.

Conclusion

MSCs are the most popular kind of stem cells used in the clinic today. Given their unique characteristics and functions, MSCs can be used in allogeneic stem cell transplantation that does not require any HLA matching, and they have been produced as off-the-shelf stem cell products (or stem cell drugs) for various clinical applications. An increasing number of off-the-shelf MSC products have been approved in different countries. Indeed, nowadays, off-the-shelf MSC technology has garnered great interest. MSCs from bone marrow, adipose tissue, umbilical cord blood, and umbilical cord tissue have been used to develop off-the-shelf products. Although there are different protocols, off-the-shelf MSC technology requires GMP-compliant conditions during cell isolation, expansion, and cryopreservation. Scale-up cell expansion is one of the most important steps in off-the-shelf MSC technology. MSCs can be scaled up in T-flasks, hollow fiber-based culture systems, or microcarrier-based culture systems. The medium for MSC production should be a xeno-free medium or serum-free medium. Human platelet lysate, human serum, or platelet-rich plasma is a suitable replacement for fetal bovine serum in cell expansion, and they help maintain MSC characteristics. Finally, the quality and safety of MSCs should be controlled to ensure that these expanded MSCs are safe and efficient for therapeutic use in the clinic.

Acknowledgment This research was partly funded by Ministry of Science and Technology, Vietnam under grant number DM.10.DA/15; by Fostering Innovation through Research, Science and Technology via project 15/FIRST/2a/SCI.

References

Anjos-Afonso F, Bonnet D (2006) Nonhematopoietic/endothelial SSEA-1+ cells define the most primitive progenitors in the adult murine bone marrow mesenchymal compartment. Blood 109(3):1298–1306. https://doi.org/10.1182/blood-2006-06-030551

Ansboro S, Roelofs AJ, De Bari C (2017) Mesenchymal stem cells for the management of rheumatoid arthritis. Curr Opin Rheumatol 29(2):201–207. https://doi.org/10.1097/bor.0000000000000370

Beeravolu N, McKee C, Alamri A, Mikhael S, Brown C, Perez-Cruet M, Rasul Chaudhry G (2017) Isolation and characterization of mesenchymal stromal cells from human umbilical cord and fetal placenta. J Vis Exp 122:PMID:28447991. https://doi.org/10.3791/55224

Berglund AK, Fortier LA, Antczak DF, Schnabel LV (2017) Immunoprivileged no more: measuring the immunogenicity of allogeneic adult mesenchymal stem cells. Stem Cell Res Ther 8(1):288. https://doi.org/10.1186/s13287-017-0742-8

Bernacki SH, Wall ME, Loboa EG (2008) Isolation of human mesenchymal stem cells from bone and adipose tissue. Methods Cell Biol 86:257–278. https://doi.org/10.1016/s0091-679x(08)00011-3

Bernardo ME, Avanzini MA, Perotti C, Cometa AM, Moretta A, Lenta E, Del Fante C et al (2007) Optimization of in vitro expansion of human multipotent mesenchymal stromal cells for cell-therapy approaches: further insights in the search for a fetal calf serum substitute. J Cell Physiol 211(1):121–130. https://doi.org/10.1002/jcp.20911

Bieback K (2004) Critical parameters for the isolation of mesenchymal stem cells from umbilical cord blood. Stem Cells 22(4):625–634. https://doi.org/10.1634/stemcells.22-4-625

Bieback K, Hecker A, Kocaömer A, Lannert H, Schallmoser K, Strunk D, Klüter H (2009) Human alternatives to fetal bovine serum for the expansion of mesenchymal stromal cells from bone marrow. Stem Cells 27(9):2331–2341. https://doi.org/10.1002/stem.139

Bifari F, Lisi V, Mimiola E, Pasini A, Krampera M (2008) Immune modulation by mesenchymal stem cells. Transfus Med Hemother 35(3):194–204. https://doi.org/10.1159/000128968

Blázquez-Prunera A, Díez JM, Gajardo R, Grancha S (2017) Human mesenchymal stem cells maintain their phenotype multipotentiality, and genetic stability when cultured using a defined xeno-free human plasma fraction. Stem Cell Res Ther 8:103. https://doi.org/10.1186/s13287-017-0552-z

Boquest AC, Shahdadfar A, Brinchmann JE, Collas P (2006) Isolation of stromal stem cells from human adipose tissue. Methods Mol Biol 325:35–46. https://doi.org/10.1385/1-59745-005-7:35

Caplan AI (1991) Mesenchymal stem cells. J Orthop Res 9:641–650

Carmelo JG, Fernandes-Platzgummer A, Diogo MM, da Silva CL, Cabral JMS (2015) A xeno-free microcarrier-based stirred culture system for the scalable expansion of human mesenchymal stem/stromal cells isolated from bone marrow and adipose tissue. Biotechnol J 10(8):1235–1247. https://doi.org/10.1002/biot.201400586

Carter RA, Wicks IP (2001) Vascular cell adhesion molecule 1 (CD106): a multifaceted regulator of joint inflammation. Arthritis Rheum 44(5):985–994. https://doi.org/10.1002/1529-0131(200105)44:5<985::aid-anr176>3.0.co;2-p

Davies JE, Walker JT, Keating A (2017) Concise review: Wharton's Jelly: the rich, but enigmatic, source of mesenchymal stromal cells. Stem Cells Transl Med 6:1620–1630

DelaRosa O, Lombardo E, Beraza A, Mancheño-Corvo P, Ramirez C, Menta R, Rico L et al (2009) Requirement of IFN-γ-mediated indoleamine 2,3-dioxygenase expression in the modulation of lymphocyte proliferation by human adiposederived stem cells. Tissue Eng A 15(10):2795–2806. https://doi.org/10.1089/ten.tea.2008.0630

Divya MS, Roshin GE, Divya TS, Rasheed V, Santhoshkumar TR, Elizabeth KE, James J, Pillai RM (2012) Umbilical cord blood-derived mesenchymal stem cells consist of a unique population of progenitors co-expressing mesenchymal stem cell and neuronal markers capable of instantaneous neuronal differentiation. Stem Cell Res Ther 3(6):57. https://doi.org/10.1186/scrt148

Dominici M, Le Blanc K, Mueller I, Slaper-Cortenbach I, Marini F, Krause D, Deans R, Keating A, Prockop D, Horwitz E (2006) Minimal criteria for defining multipotent mesenchymal stromal cells. the international society for cellular therapy position statement. Cytotherapy 8:315–317

dos Santos F, Andrade PZ, Abecasis MM, Gimble JM, Chase LG, Campbell AM, Boucher S, Vemuri MC, da Silva CL, Cabral JMS (2011) Toward a clinical grade expansion of mesenchymal stem cells from human sources: a microcarrier-based culture system under xeno-free conditions. Tissue Eng C Methods 17(12):1201–1210. https://doi.org/10.1089/ten.tec.2011.0255

dos Santos F, Campbell A, Fernandes-Platzgummer A, Andrade PZ, Gimble JM, Wen Y, Boucher S, Vemuri MC, da Silva CL, Cabral JMS (2014) A xenogeneic-free bioreactor system for the clinical-scale expansion of human mesenchymal stem/stromal cells. Biotechnol Bioeng 111(6):1116–1127. https://doi.org/10.1002/bit.25187

Du YM, Zhuansun YX, Chen R, Lin L, Lin Y, Li JG (2018) Mesenchymal stem cell exosomes promote immunosuppression of regulatory T cells in asthma. Exp Cell Res 363:114–120

Ducret M, Fabre H, Degoult O, Atzeni G, McGuckin C, Forraz N, Mallein-Gerrin F, Perrier-Groult E, Fargues JC (2016) A standardized procedure to obtain mesenchymal stem/stromal cells from minimally manipulated dental pulp and Wharton's Jelly samples. Bull Group Int Rech Sci Stomatol Odontol 53:e37

English K, Ryan JM, Tobin L, Murphy MJ, Barry FP, Mahon BP (2009) Cell contact prostaglandin E2and transforming growth factor beta 1 play non-redundant roles in human mesenchymal stem cell induction of CD4+CD25 highforkhead box P3+ regulatory T cells. Clin Exp Immunol 156(1):149–160. https://doi.org/10.1111/j.1365-2249.2009.03874.x

Escobar CH, Chaparro O (2016) Xeno-free extraction culture, and cryopreservation of human adipose-derived mesenchymal stem cells. Stem Cells Transl Med 5(3):358–365. https://doi.org/10.5966/sctm.2015-0094

Esmaeli A, Moshrefi M, Shamsara A, Eftekhar-Vaghefi SH, Nematollahi-Mahani SN (2016) Xeno-free culture condition for human bone marrow and umbilical cord matrix-derived mesenchymal stem/stromal cells using human umbilical cord blood serum. Int J Reprod Biomed 14:567–576

Falcon-Girard K, Fuhrmann A, Briddell R, Walters M, Foster K, Kraus M (2013) Characterization of umbilical cord blood (UCB) and umbilical cord tissue (UCT) stem cells from premature infants. Cytotherapy 15(4):S21. https://doi.org/10.1016/j.jcyt.2013.01.077

Fekete N, Gadelorge M, Fürst D, Maurer C, Dausend J, Fleury-Cappellesso S, Mailänder V et al (2012) Platelet lysate from whole blood-derived pooled platelet concentrates and apheresis-derived platelet concentrates for the isolation and expansion of human bone marrow mesenchymal stromal cells: production process content and identification of active components. Cytotherapy 14(5):540–554. https://doi.org/10.3109/14653249.2012.655420

Fibbe WE, Nauta AJ, Roelofs H (2007) Modulation of immune responses by mesenchymal stem cells. Annals of the New York Academy of Sciences 1106(1):272–278. https://doi.org/10.1196/annals.1392.025

Francis MP, Sachs PC, Elmore LW, Holt SE (2010) Isolating adipose-derived mesenchymal stem cells from lipoaspirate blood and saline fraction. Organogenesis 6:11–14

Friedenstein AJ, Chailakhjan RK, Lalykina KS (1970) The development of fibroblast colonies in monolayer cultures of guinea-pig bone marrow and spleen cells. Cell Tissue Kinet 3:393–403

Friedenstein AJ, Deriglasova UF, Kulagina NN, Panasuk AF, Rudakowa SF, Luriá EA, Ruadkow IA (1974) Precursors for fibroblasts in different populations of hematopoietic cells as detected by the in vitro colony assay method. Exp Hematol 2:83–92

Fujii S, Miura Y, Iwasa M, Yoshioka S, Fujishiro A, Sugino N, Kaneko H et al (2017) Isolation of mesenchymal stromal/stem cells from cryopreserved umbilical cord blood cells. J Clin Exp Hematop 57:1–8

Gang EJ, Bosnakovski D, Figueiredo CA, Visser JW, Perlingeiro RCR (2007) SSEA-4 identifies mesenchymal stem cells from bone marrow. Blood 109(4):1743–1751. https://doi.org/10.1182/blood-2005-11-010504

Gennery A (2016) Faculty of 1000 evaluation for efficacy and safety of ex vivo cultured adult human mesenchymal stem cells (prochymal) in pediatric patients with severe refractory acute graft-versus-host disease in a compassionate use study. Faculty of 1000 Ltd, London. https://doi.org/10.3410/f.720537847.793519788

Gersh BJ (2010) A randomized double-blind, placebo-controlled, dose-escalation study of intravenous adult human mesenchymal stem cells (prochymal) after acute myocardial infarction. Yearb Cardiol 2010:389–391. https://doi.org/10.1016/s0145-4145(10)79814-2

Gottipamula S, Muttigi MS, Chaansa S, Ashwin KM, Priya N, Kolkundkar U, Raj SS, Majumdar AS, Seetharam RN (2013) Large-scale expansion of pre-isolated bone marrow mesenchymal stromal cells in serum-free conditions. J Tissue Eng Regen Med 10(2):108–119. https://doi.org/10.1002/term.1713

Goujon E. (1869) Recherches expérimentales sur les propriétés physiologiques de la moelle des os. J Anat Physiol 6:399–412.

Gottipamula S, Ashwin KM, Muttigi MS, Kannan S, Kolkundkar U, Seetharam RN (2014) Isolation expansion and characterization of bone marrow-derived mesenchymal stromal cells in serum-free conditions. Cell Tissue Res 356(1):123–135. https://doi.org/10.1007/s00441-013-1783-7

Haack-Sørensen M, Follin B, Juhl M, Brorsen SK, Søndergaard RH, Kastrup J, Ekblond A (2016) Culture expansion of adipose derived stromal cells. A closed automated quantum cell expansion system compared with manual flask-based culture. J Transl Med 14(1):319. https://doi.org/10.1186/s12967-016-1080-9

Haack-Sørensen M, Juhl M, Follin B, Søndergaard RH, Kirchhoff M, Kastrup J, Ekblond A (2018) Development of large-scale manufacturing of adipose-derived stromal cells for clinical applications using bioreactors and human platelet lysate. Scand J Clin Lab Investig 78(4):293–300. https://doi.org/10.1080/00365513.2018.1462082

Han KH, Ro H, Hong JH, Lee EM, Cho B, Yeom HJ, Kim M-G, Kook-Hwan O, Ahn C, Yang J (2011) Immunosuppressive mechanisms of embryonic stem cells and mesenchymal stem cells in alloimmune response. Transplant Immunol 25(1):7–15. https://doi.org/10.1016/j.trim.2011.05.004

Han Y-F, Tao R, Sun T-J, Chai J-K, Xu G, Liu J (2013) Optimization of human umbilical cord mesenchymal stem cell isolation and culture methods. Cytotechnology 65(5):819–827. https://doi.org/10.1007/s10616-012-9528-0

Harris D (2013) Umbilical cord tissue mesenchymal stem cells: characterization and clinical applications. Curr Stem Cell Res Ther 8(5):394–399. https://doi.org/10.2174/1574888x11308050006

Hartmann I, Hollweck T, Haffner S, Krebs M, Meiser B, Reichart B, Eissner G (2010) Umbilical cord tissue-derived mesenchymal stem cells grow best under GMP-compliant culture conditions and maintain their phenotypic and functional properties. J Immunol Methods 363(1):80–89. https://doi.org/10.1016/j.jim.2010.10.008

Hassan G, Kasem I, Soukkarieh C, Aljamali M (2017) A simple method to isolate and expand human umbilical cord derived mesenchymal stem cells: using explant method and umbilical cord blood serum. Int J Stem Cells 10(2):184–192. https://doi.org/10.15283/ijsc17028

Hatlapatka T, Moretti P, Lavrentieva A, Hass R, Marquardt N, Jacobs R, Kasper C (2011) Optimization of culture conditions for the expansion of umbilical cord-derived mesenchymal stem or stromal cell-like cells using xeno-free culture conditions. Tissue Eng C Methods 17(4):485–493. https://doi.org/10.1089/ten.tec.2010.0406

Hildebrandt C, Büth H, Thielecke H (2009) Influence of cell culture media conditions on the osteogenic differentiation of cord blood-derived mesenchymal stem cells. Ann Anat 191(1):23–32. https://doi.org/10.1016/j.aanat.2008.09.009

Hong HS, Kim YH, Son Y (2012) Perspectives on mesenchymal stem cells: tissue repair immune modulation, and tumor homing. Arch Pharm Res 35(2):201–211. https://doi.org/10.1007/s12272-012-0201-0

Horn P, Bork S, Diehlmann A, Walenda T, Eckstein V, Ho A, Wagner W (2008) Isolation of human mesenchymal stromal cells is more efficient by red blood cell lysis. Cytotherapy 10(7):676–685. https://doi.org/10.1080/14653240802398845

Horn P, Bork S, Wagner W (2011) Standardized isolation of human mesenchymal stromal cells with red blood cell lysis. In: Vemuri MC (ed) Mesenchymal stem cell assays and applications. Humana, Louisville, KY, pp 23–35. https://doi.org/10.1007/978-1-60761-999-4_3

Jarvinen L, Badri L, Wettlaufer S, Ohtsuka T, Standiford TJ, Toews GB, Pinsky DJ, Peters-Golden M, Lama VN (2008) Lung resident mesenchymal stem cells isolated from human lung allografts inhibit T cell proliferation via a soluble mediator. The Journal of Immunology 181(6):4389–4396. https://doi.org/10.4049/jimmunol.181.6.4389

Jones M, Varella-Garcia M, Skokan M, Bryce S, Schowinsky J, Peters R, Vang B et al (2013) Genetic stability of bone marrow-derived human mesenchymal stromal cells in the quantum system. Cytotherapy 15(11):1323–1339. https://doi.org/10.1016/j.jcyt.2013.05.024

Jung YH, Lee S-J, Sang Yub O, Lee HJ, Ryu JM, Han HJ (2015) Oleic acid enhances the motility of umbilical cord blood derived mesenchymal stem cells through EphB2-dependent F-actin formation. Biochim Biophys Acta 1853(8):1905–1917. https://doi.org/10.1016/j.bbamcr.2015.05.006

Kargozar S, Mozafari M, Hashemian SJ, Brouki MP, Hamzehlou S, Soleimani M, Joghataei MT et al (2018) Osteogenic potential of stem cells-seeded bioactive nanocomposite scaffolds: a comparative study between human mesenchymal stem cells derived from bone, umbilical cord Wharton's Jelly, and adipose tissue. J Biomed Mater Res B Appl Biomater 106:61–72

Kastrup J, Haack-Sørensen M, Juhl M, Søndergaard RH, Follin B, Lund LD, Johansen EM et al (2017) Cryopreserved off-the-shelf allogeneic adipose-derived stromal cells for therapy in patients with ischemic heart disease and heart failure—a safety study. Stem Cells Transl Med 6(11):1963–1971. https://doi.org/10.1002/sctm.17-0040

Kim ES, Jeon HB, Lim H, Shin JH, Park SJ, Jo YK, Wonil O, Yang YS, Cho D-H, Kim J-Y (2015) Conditioned media from human umbilical cord blood-derived mesenchymal stem cells inhibits melanogenesis by promoting proteasomal degradation of MITF. PLoS One 10(5):e0128078. https://doi.org/10.1371/journal.pone.0128078

Knaän-Shanzer S (2014) Concise review: the immune status of mesenchymal stem cells and its relevance for therapeutic application. Stem Cells 32(3):603–608. https://doi.org/10.1002/stem.1568

Koç ON, Gerson SL (2003) Mesenchymal stem cells in allogeneic transplantation. In: Allogeneic stem cell transplantation. Leukemia & Lymphoma Society, New York City, NY, pp 151–158. https://doi.org/10.1007/978-1-59259-333-0_11

Kocaoemer A, Kern S, Klüter H, Bieback K (2007) Human AB serum and thrombin-activated platelet-rich plasma are suitable alternatives to fetal calf serum for the expansion of mesenchymal stem cells from adipose tissue. Stem Cells 25(5):1270–1278. https://doi.org/10.1634/stemcells.2006-0627

Kolf CM, Cho E, Tuan RS (2007) Mesenchymal stromal cells. Biology of adult mesenchymal stem cells: regulation of niche, self-renewal and differentiation. Arthritis Res Ther 9(1):204. https://doi.org/10.1186/ar2116

Krampera M (2002) Bone marrow mesenchymal stem cells inhibit the response of naive and memory antigen-specific T cells to their cognate peptide. Blood 101(9):3722–3729. https://doi.org/10.1182/blood-2002-07-2104

Krampera M, Cosmi L, Angeli R, Pasini A, Liotta F, Andreini A, Santarlasci V et al (2006) Role for interferon-γ in the immunomodulatory activity of human bone marrow mesenchymal stem cells. Stem Cells 24(2):386–398. https://doi.org/10.1634/stemcells.2005-0008

Kurtzberg J, Prasad V, Grimley MS, Horn B, Carpenter PA, Jacobsohn D, Prockop S (2010) Allogeneic human mesenchymal stem cell therapy (prochymal) as a rescue agent for severe treatment resistant GVHD in pediatric patients. Biol Blood Marrow Transplant 16(2):S169. https://doi.org/10.1016/j.bbmt.2009.12.056

Kurtzberg J, Prockop S, Teira P, Bittencourt H, Lewis V, Chan KW, Horn B et al (2014) Allogeneic human mesenchymal stem cell therapy (remestemcel-L prochymal) as a rescue agent for severe refractory acute graft-versus-host disease in pediatric patients. Biol Blood Marrow Transplant 20(2):229–235. https://doi.org/10.1016/j.bbmt.2013.11.001

Lange C, Cakiroglu F, Spiess A-N, Cappallo-Obermann H, Dierlamm J, Zander AR (2007) Accelerated and safe expansion of human mesenchymal stromal cells in animal serum-free medium for transplantation and regenerative medicine. J Cell Physiol 213(1):18–26. https://doi.org/10.1002/jcp.21081

Lawson T, Kehoe DE, Schnitzler AC, Rapiejko PJ, Der KA, Philbrick K, Punreddy S et al (2017) Process development for expansion of human mesenchymal stromal cells in a 50L single-use stirred tank bioreactor. Biochem Eng J 120:49–62. https://doi.org/10.1016/j.bej.2016.11.020

Le PT-B, Duong TM, Ngoc Bich V, Van Pham P (2016) Umbilical cord derived stem cell (ModulatistTM) transplantation for severe chronic obstructive pulmonary disease: a report of two cases. Biomed Res Ther 3(10):49. https://doi.org/10.7603/s40730-016-0049-x

Lee OK (2004) Isolation of multipotent mesenchymal stem cells from umbilical cord blood. Blood 103(5):1669–1675. https://doi.org/10.1182/blood-2003-05-1670

Lee MW, Choi J, Yang MS, Moon YJ, Park JS, Kim HC, Kim YJ (2004) Mesenchymal stem cells from cryopreserved human umbilical cord blood. Biochem Biophys Res Commun 320(1):273–278. https://doi.org/10.1016/j.bbrc.2004.04.206

Lee M-S, Youn C, Kim J, Park B, Ahn J, Hong S, Kim Y-D, Shin Y, Park S (2017) Enhanced cell growth of adipocyte-derived mesenchymal stem cells using chemically-defined serum-free media. Int J Mol Sci 18(8):1779. https://doi.org/10.3390/ijms18081779

Li FR, Wang XG, Deng CY, Qi H, Ren LL, Zhou HX (2010) Immune modulation of co-transplantation mesenchymal stem cells with islet on T and dendritic cells. Clin Exp Immunol 161(2):357–363. https://doi.org/10.1111/j.1365-2249.2010.04178.x

Lindroos B, Boucher S, Chase L, Kuokkanen H, Huhtala H, Haataja R, Vemuri M, Suuronen R, Miettinen S (2009) Serum-free xeno-free culture media maintain the proliferation rate and multipotentiality of adipose stem cells in vitro. Cytotherapy 11(7):958–972. https://doi.org/10.3109/14653240903233081

Liu C-H, Wu M-L, Hwang S-M (2007) Optimization of serum free medium for cord blood mesenchymal stem cells. Biochem Eng J 33(1):1–9. https://doi.org/10.1016/j.bej.2006.08.005

Martin PJ, Uberti JP, Soiffer RJ, Klingemann H, Waller EK, Daly AS, Herrmann RP, Kebriaei P (2010) Prochymal improves response rates in patients with steroid-refractory acute graft versus host disease (SR-GVHD) involving the liver and gut: results of a randomized placebo-controlled, multicenter phase III trial in GVHD. Biol Blood Marrow Transplant 16(2):S169–S170. https://doi.org/10.1016/j.bbmt.2009.12.057

Mizukami A, Fernandes-Platzgummer A, Carmelo JG, Swiech K, Covas DT, Cabral JMS, da Silva CL (2016) Stirred tank bioreactor culture combined with serum-/xenogeneic-free culture medium enables an efficient expansion of umbilical cord-derived mesenchymal stem/stromal cells. Biotechnol J 11(8):1048–1059. https://doi.org/10.1002/biot.201500532

Najima Y, Ohashi K (2017) Mesenchymal stem cells: the first approved stem cell drug in japan. J Hematopoietic Cell Transplant 6(3):125–132. https://doi.org/10.7889/hct-16-031

Nauta AJ (2006) Donor-derived mesenchymal stem cells are immunogenic in an allogeneic host and stimulate donor graft rejection in a nonmyeloablative setting. Blood 108(6):2114–2120. https://doi.org/10.1182/blood-2005-11-011650

Nold P, Brendel C, Neubauer A, Bein G, Hackstein H (2013) Good manufacturing practice-compliant animal-free expansion of human bone marrow derived mesenchymal stroma cells in a closed hollow-fiber-based bioreactor. Biochem Biophys Res Commun 430(1):325–330. https://doi.org/10.1016/j.bbrc.2012.11.001

O'Connor SL, Sepulveda CA, Kaur I, Sumari RD, McMannis JD (2007) Characterization of BioSafe SEPAX manufactured stem cell intended for cardiac cell therapy. Blood 110:4064

Owen M (1988) Marrow stromal stem cells. J Cell Sci Suppl 10:63–76

Park YB, Ha CW, Lee CH, Yoon YC, Park YG (2017) Cartilage regeneration in osteoarthritic patients by a composite of allogeneic umbilical cord blood-derived mesenchymal stem cells and hyaluronate hydrogel: results from a clinical trial for safety and proof-of-concept with 7 years of extended follow-up. Stem Cells Transl Med 6:613–621

Paula ACC, Martins TMM, Zonari A, Frade SPPJ, Angelo PC, Gomes DA, Goes AM (2015) Human adipose tissue-derived stem cells cultured in xeno-free culture condition enhance c-MYC expression increasing proliferation but bypassing spontaneous cell transformation. Stem Cell Res Ther 6(1). https://doi.org/10.1186/s13287-015-0030-4

Pérez-Ilzarbe M, Díez-Campelo M, Aranda P, Tabera S, Lopez T, del Cañizo C, Merino J et al (2009) Comparison of ex vivo expansion culture conditions of mesenchymal stem cells for human cell therapy. Transfusion 49(9):1901–1910. https://doi.org/10.1111/j.1537-2995.2009.02226.x

Pharmacopoeia [EP] (2005) Sterility. Biological tests. In: EUROPEAN PHARMACOPOEIA 5.0, pp 145–148

Pharmacopoeia [USP] (2011) Sterility test. In: US PHARMACOPOEIA USP29. http://pharmacopeia.cn/v29240/usp29nf24s0_c71.html

Pieper IL, Smith R, Bishop JC, Aldalati O, Chase AJ, Morgan G, Thornton CA (2017) Isolation of mesenchymal stromal cells from peripheral blood of ST elevation myocardial infarction patients. Artif Organs 41:654–666

Pittenger MF (1999) Multilineage potential of adult human mesenchymal stem cells. Science 284(5411):143–147. https://doi.org/10.1126/science.284.5411.143

Pranke P, Canabarro R (2008) Stem cells from umbilical cord blood. Front Cord Blood Sci. Springer, London, pp 27–90. https://doi.org/10.1007/978-1-84800-167-1_3

Prasad VK, Lucas KG, Kleiner GI, Talano JAM, Jacobsohn D, Broadwater G, Monroy R, Kurtzberg J (2011) Efficacy and safety of ex vivo cultured adult human mesenchymal stem cells (prochymal) in pediatric patients with severe refractory acute graft-versus-host disease in a compassionate use study. Biol Blood Marrow Transplant 17(4):534–541. https://doi.org/10.1016/j.bbmt.2010.04.014

Prockop DJ, Sekiya I, Colter DC (2001) Isolation and characterization of rapidly self-renewing stem cells from cultures of human marrow stromal cells. Cytotherapy 3(5):393–396. https://doi.org/10.1080/146532401753277229

Pytlík R, Stehlík D, Soukup T, Kalbáčová M, Rypáček F, Trč T, Mulinková K et al (2009) The cultivation of human multipotent mesenchymal stromal cells in clinical grade medium for bone tissue engineering. Biomaterials 30(20):3415–3427. https://doi.org/10.1016/j.biomaterials.2009.03.001

Rajala K, Lindroos B, Hussein SM, Lappalainen RS, Pekkanen-Mattila M, Inzunza J, Rozell B et al (2010) A defined and xeno-free culture method enabling the establishment of clinical-grade human embryonic induced pluripotent and adipose stem cells. PLoS ONE 5(4):e10246. https://doi.org/10.1371/journal.pone.0010246

Ranjbaran H, Abediankenari S, Mohammadi M, Jafari N, Khalilian A, Rahmani Z, Momeninezhad AM, Ebrahimi P (2018) Wharton's Jelly derived-mesenchymal stem cells: isolation and characterization. Acta Med Iran 56:28–33

Rojewski MT, Fekete N, Baila S, Nguyen K, Fürst D, Antwiler D, Dausend J et al (2013) GMP-compliant isolation and expansion of bone marrow-derived MSCs in the closed automated device quantum cell expansion system. Cell Transplant 22(11):1981–2000. https://doi.org/10.3727/096368912x657990

Ryan JM, Barry F, Murphy JM, Mahon BP (2007) Interferon-γ does not break but promotes the immunosuppressive capacity of adult human mesenchymal stem cells. Clin Exp Immunol 149(2):353–363. https://doi.org/10.1111/j.1365-2249.2007.03422.x

Schofield R (1978) The relationship between the spleen colony-forming cell and the haemopoietic stem cell. Blood Cells 4:7–25

Semedo P, Correa-Costa M, Cenedeze MA, Malheiros DMAC, dos Reis MA, Shimizu MH, Seguro AC, Pacheco-Silva A, Câmara NOS (2009) Mesenchymal stem cells attenuate renal fibrosis through immune modulation and remodeling properties in a rat remnant kidney model. Stem Cells 27(12):3063–3073. https://doi.org/10.1002/stem.214

Sheridan C (2018) First off-the-shelf mesenchymal stem cell therapy nears European approval. Nat Biotechnol 36:212–214

Shetty P, Bharucha K, Tanavde V (2007) Human umbilical cord blood serum can replace fetal bovine serum in the culture of mesenchymal stem cells. Cell Biol Int 31(3):293–298. https://doi.org/10.1016/j.cellbi.2006.11.010

Smith J, Robert KP, Petry F, Powell N, Delzeit J, Weiss ML (2016) Standardizing umbilical cord mesenchymal stromal cells for translation to clinical use: selection of GMP-compliant medium and a simplified isolation method. Stem Cells Int 2016:1–14. https://doi.org/10.1155/2016/6810980

Tavassoli M, Crosby WH (1968) Transplantation of marrow to extramedullary sites. Science, 161(3836), pp.54–56.

Tzouvelekis A, Koliakos G, Ntolios P, Baira I, Bouros E, Oikonomou A, Zissimopoulos A et al (2011) Stem cell therapy for idiopathic pulmonary fibrosis: a protocol proposal. J Transl Med 9(1):182. https://doi.org/10.1186/1479-5876-9-182

Van Pham P, Phan NK (2014) Production of good manufacturing practice-grade human umbilical cord blood-derived mesenchymal stem cells for therapeutic use. Methods Mol Biol 1283:73–85. https://doi.org/10.1007/7651_2014_125

Van Pham P, Bui KH-T, Ngo DQ, Khuat LT, Phan NK (2013) Transplantation of nonexpanded adipose stromal vascular fraction and platelet-rich plasma for articular cartilage injury treatment in mice model. J Med Eng 2013:1–7. https://doi.org/10.1155/2013/832396

Van Pham P, Phan NL-C, Le DM, Le PT-B, Tran TD-X, Phan NK (2014a) Good manufacturing practice-compliant isolation and culture of human bone marrow mesenchymal stem cells. Progr Stem Cell 1:56. https://doi.org/10.15419/psc.v1i01.117

Van Pham P, Ngoc Bich V, Pham VM, Truong NH, Pham TL-B, Dang LT-T, Nguyen TT, Bui AN-T, Phan NK (2014b) Good manufacturing practice-compliant isolation and culture of human umbilical cord blood-derived mesenchymal stem cells. J Transl Med 12(1):56. https://doi.org/10.1186/1479-5876-12-56

Van Pham P, Ngoc Bich V, Phan NL-C, Le DM, Truong NC, Truong NH, Bui KH-T, Phan NK (2014c) Good manufacturing practice-compliant isolation and culture of human adipose derived stem cells. Biomed Res Ther 1(4). https://doi.org/10.7603/s40730-014-0021-6

Van Pham P, Truong NC, Le PT-B, Tran TD-X, Ngoc Bich V, Bui KH-T, Phan NK (2015) Isolation and proliferation of umbilical cord tissue derived mesenchymal stem cells for clinical applications. Cell Tissue Bank 17(2):289–302. https://doi.org/10.1007/s10561-015-9541-6

Van Pham P, Bich NV, Phan NK (2016) Umbilical cord-derived stem cells (modulatisttm) show strong immunomodulation capacity compared to adipose tissue-derived or bone marrow-derived mesenchymal stem cells. Biomed Res Ther 3(6). https://doi.org/10.7603/s40730-016-0029-1

Van PP, Truong NC, Le PT, Tran TD, Vu NB, Bui KH, Phan NK (2016) Isolation and proliferation of umbilical cord tissue derived mesenchymal stem cells for clinical applications. Cell Tissue Bank 17:289–302

van Vollenstee FA, Hoffmann D, Pepper MS (2016) Harvesting and collection of adipose tissue for the isolation of adipose-derived stromal/stem cells. In: Van Pham P (ed) Stem cells in clinical applications. Springer, New York, NY, pp 199–220. https://doi.org/10.1007/978-3-319-40073-0_10

Wu X, Kang H, Liu X, Gao J, Zhao K, Ma Z (2016) Serum and xeno-free chemically defined, no-plate-coating-based culture system for mesenchymal stromal cells from the umbilical cord. Cell Proliferation 49(5):579–588. https://doi.org/10.1111/cpr.12279

Yoo SW, Chang DY, Lee HS, Kim GH, Park JS, Ryu BY, Joe EH, Lee YD, Kim SS, Suh-Kim H (2013) Immune following suppression mesenchymal stem cell transplantation in the ischemic brain is mediated by TGF-β. Neurobiol Dis 58:249–257

Zimmerlin L, Donnenberg VS, Pfeifer ME, Michael Meyer E, Bruno P, Peter Rubin J, Donnenberg AD (2009) Stromal vascular progenitors in adult human adipose tissue. Cytometry Part A 77(1):22–30. https://doi.org/10.1002/cyto.a.20813

Chapter 8
Ethical and Legal Issues of Cord Blood Stem Cell Banking

Luciana Riva, Giovanna Floridia, and Carlo Petrini

Introduction

Umbilical cord blood (UBC), once regarded as discarded biological material, is today a precious resource in the clinical practice because it is rich in transplantable hematopoietic stem cells (HPCs). Its cellular composition is very similar to that of bone marrow and includes hematopoietic stem cells (HSCs), progenitor cells (HPCs, CD34+, and CD133+ cells), and mesenchymal stem cells (MSCs). UCB cell transplantation (UCBT) is today routinely used for the treatment of a range of malignant and nonmalignant hematologic disorders (e.g., leukemia, immune deficiencies, and congenital disorders). By 2013, more than 30,000 hematopoietic stem cell transplants (HSCT) worldwide have been performed in different malignancies and disorders using cord blood as the source of stem cells (Shearer et al. 2017).

History

About 40 years ago it was first suggested that stem and progenitor cells were present in human cord blood. In the early 1980s, the possibility of using UCB as a source of transplantable HSCs and HPCs was raised by Hal Broxmeyer, who is now recognized as the founder of the field, during a meeting with Edward A. Boyse and Judith Bard (Ballen et al. 2013). The same researchers funded the company Broxmeyer, at the Indiana University School of Medicine, with a 2-year grant to study the biology and

L. Riva (✉) · G. Floridia · C. Petrini
Bioethics Unit, Istituto Superiore di Sanità (Italian National Institute of Health), Rome, Italy
e-mail: luciana.riva@iss.it; giovanna.floridia@iss.it; carlo.petrini@iss.it

P. V. Pham (ed.), *Stem Cell Drugs - A New Generation of Biopharmaceuticals*,
Stem Cells in Clinical Applications, https://doi.org/10.1007/978-3-319-99328-7_8

cryopreservation of UCB cells, and began to explore the possibility of using UCB in transplantation. These studies led to the first umbilical cord blood transplant (UCBT), which was performed in 1988, in France, in a boy suffering from Fanconi anemia (FA). The UCB was collected at the birth of his healthy sister with an identical HLA and, then, cryopreserved at the IUSM (Gluckman et al. 1989). 2018 marks the 30th anniversary of this milestone. This success paved the way for a new field of research and development. Indeed, since then, the comprehension of UCB cells biology and their potential for clinical use has increased exponentially and, therefore, unrelated CB transplantation is nowadays an established practice treatment option. It has been proved to have some significant advantages over the use of bone marrow stem cells for transplants, one of the most important being that it does not require a perfect match and therefore affords a greater flexibility for HLA matching purposes. Another important difference is that UCB units can be harvested and stored before they are needed for transplantation and the donation is not an invasive procedure. UCB donation is a purely altruistic gesture and, from a safety perspective, has a small risk to transmit communicable diseases. Banks to store umbilical CB were first introduced in the 1990s; moreover, the absence of any particular ethical concern has facilitated studies on cord blood for stem cell therapy.

UCB Banks

The first UCB bank was set up in 1991 at the New York Blood Center with the purpose of storing units for public use (allogeneic donation). Public banks collect and store donated UCB in monitored cryopreservation tanks until it is searched, matched, and distributed for any patient in need of an HSC transplant. It is possible to donate infant's UCB to a public bank if the unit meets the required criteria for banking: generally, the facility has a minimum volume (and cell number) that is considered adequate for a transplant patient and that will be accepted. Products that do not meet these requirements are discarded or used for research purposes, subject to a statement in the informed consent document. Private UCB banks, which are profit-making facilities that provide a service for families who want to store UCB stem cells privately (autologous donation) as a form of medical insurance, arose in the USA in approximately the same period. The latter market grew very rapidly. At present, the government regulations in the USA for family banks are less stringent than those for public banks in comparison with many other countries where national health authorities regulate in the same way family and public banks. In some European countries private banks are banned. In Italy and France the law does not allow to privately store one's cord blood. In Italy, for instance, cord blood banking is only authorized as a public conservation structure but a family banking program, so-called "dedicated storage," is allowed for parents who already have a child with a disease considered treatable using cord blood transplantation (that is a scientifically validated treatment).

It has been shown that in private cord blood banks, compared with public cord blood ones, there is not a full exploitation for treatment, the quality control is less regulated, and costs for the family are higher (Ballen et al. 2013). To date about 215 family UCB banks are located in 54 countries and at least 200 marketing affiliates serve over 70 countries (Ballen et al. 2015). Private cord blood banking can be expensive with a fee upon acceptance of the sample (usually between US$1500 and US$2000) plus an annual storage fee (Petrini 2014). The clinical efficacy of autologous stem cells for the purposes of prevention is statistically very low, and private storage for later autologous use does not appear in relevant documents issued by the most authoritative institutions (Petrini 2014; Petrini 2015). To the present, public cord blood banks are storing approximately 800,000 unrelated cord blood units, and private ones more than 5,000,000 (Kurtzberg 2017). Differently from private banks, public ones make samples for potential recipients available through registries at the international level. There is a global system of public cord blood banks and transplant centers connected by networks to facilitate the exchange of information. There is a link between national registries and international ones (e.g., the European Marrow Donor Information System and Bone Marrow Donor Worldwide) which allows to identify the most appropriate sample for each patient needing a transplant. At a national level, the size of each registry is of crucial importance: the larger the number of units registered, the greater the probability of finding a clinically useful match between donor and recipient. The international dimension is decisive, among other things, because it is necessary for access to immunotypes that are extensively dispersed (Petrini 2014). In many countries (including the USA) not every hospital is associated with a public bank, which means not every donor may be able to donate and the number of units for public use is considerably lower than that required. An issue to be considered is that ethnic and/or minority patients, needing a cord-blood transplant, could have more difficulties to find units with HLA match (Shearer et al. 2017). Therefore, self-sufficiency of a country depends not only on health care policies but also on the ethnic homogeneity of its population.

Recently, in Europe, a private-public mixed UCB banks model emerged. A *hybrid* UCB bank is a private institution in which cord blood units are stored for possible public or private use. Different feasible models for hybrid banks have been proposed. The British Virgin Health Bank is an example in which the units of cord blood preserved are destined in 80% for allogeneic use (in an inventory available for public use) and in 20% for autologous use (exclusively for the client). The "public" portion of the product may be donated to a potential HSCT center if the inventory was searched. Nowadays there are other hybrid models (Petrini 2014), for instance some private banks may create a partnership with universities and public institutions. As already argued this model, based on a public–private partnership, could support financially the public biobank network (Pontifical Academy for Life Banks of Umbilical Cord Blood 2013).

Regulation

While increasing banking and clinical applications, the development of a regulatory system has become necessary, particularly as a means of guaranteeing both the protection of the donor and the recipients and the quality of the products. Standards have been set worldwide to act as minimum guidelines for CB bank operations. The standards and accreditation system in CB banking ensures that processes and products are of high quality on an international level (Armitage 2016). It is mandatory that any criteria adopted for CB banking is international since CB products frequently cross international borders as the best CB unit available for a patient, selected according to HLA type and cell dose, is often likely to be located in another country. The quality of the products must be guaranteed in terms of safeness, pureness and potency, bearing in mind that they may easily have been banked over two decades before (units can be stored in the cryopreserved state for at least 20 years without harming the viability of the cells) (Ballen et al. 2013).

The most important international accreditation standards for umbilical cord blood bank operations have been designed by the American Association of Blood Banks (AABB) and by the Foundation for the Accreditation of Cellular Therapy together with NetCord (NetCord-FACT 2015). There are, however, other international organizations, such as the Joint Accreditation Committee ICST (International Society for Cellular Therapy) and EBMT (European Group for Blood and Marrow transplant), International Organization for Standardization (ISO) and International Society of Blood Transfusion (ISBT), that offer accreditation related to registry operations of HSC. Furthermore, the World Marrow Donor Association (WMDA) standards aim to improve the quality of unrelated hematopoietic stem cell donor registries. Through the process of accreditation, the bank proves to be in line with the most up-to-date standards. At the individual state level, competent authorities may also regulate products within their country. Each National registry must therefore follow any additional law, regulation, practice and procedure that apply in that particular nation. For example individual states may regulate biological products differently (e.g., as biological drugs, blood or tissue products). Up-to-date private banks store a higher number of UCB compared to public ones depending also on the adoption of less stringent criteria for acceptability (e.g., number of cells, viability) with the risk of a not optimal quality of the unit (Petrini 2014).

Ethical issues related to collection, storage and use of CB stem cells have been addressed in several and different documents issued by national and international institutions (Petrini 2013). The analysis made by Petrini showed a considerable production of documents by National Bioethics Committee especially in Europe. Eight of the 27 member countries belonging to the European Union have produced a total of 11 official statements. Most of these document where published after the release of the Opinion 19 "Ethical Aspects of Umbilical CB Banking" by the European Commission's Group on Ethics in Science and New Technologies. The group highlighted some basic ethical principles to be considered: "the principles of justice and solidarity, as regards to fair access to healthcare services; the principle of beneficence, or the obligation to do good, especially in the area of health care; the principle of nonmaleficence, or the obligation not to harm, including the obligation to protect vulnerable groups and individuals, to respect privacy and confidentiality; the principle of proportionality which implies a balance between means and objectives."

Besides the above mentioned principles it is necessary to consider the principle of respect for human dignity and integrity, the right to self-determination on the basis of full and correct information and the prohibition of a financial gain from the human body and its parts (art. 21 Convention of Human Rights and Biomedicine). It is noteworthy that in the USA most commentaries have been emanated from professional organizations (Petrini 2013).

Private vs. Public Bank

The ethical issue of public vs. private banks has been widely debated. Private banks treat the biological resource as a product with an economic value, inserted within a market model of the supply and demand type. Anyone who is opposed to the concept of gaining an economic advantage from the body and its parts encourages social solidarity between citizens and an approach based on voluntary and free donations. From this perspective cord blood, as in the same way blood or organs, is considered a "common good" whose availability responds to a health need of any citizen. Besides attention to operating standards, a continuing ethical scrutiny, particularly in view of the rapid growth of the market, is basically needed to ensure that the development of the banking system serves the common good. In the last decade several bodies and public authorities worldwide expressed the opinion that private cord blood banking has not any real clinical application with the exception of some specific cases. In case of need, furthermore, thanks to the public international network, anyone in need has the opportunity to obtain, through the international circuit, the most suitable sample for transplant. Support for the private bank is generally based on the principle of respect for individual autonomy; patients are considered as "rational agents with the rights to be informed and to make choices which affect themselves and their offspring" and consequently anyone should have the right to save cells for future use at their own expense. The crucial issues here are: how futuristic this future use is and to what extent promises, made to families about the potential future use, may justify private storage with the associated costs Private UCB banks advertise the service they offer defining the autologous storage as a "biological insurance." The individual autonomy and the freedom of choice fill up with value when there is no misunderstanding or a false expectation. The narrative of a possible utility of UCB-derived stem cells in a contest of regenerative medicine, especially, should not be designed to attract business.

Bioeconomic Aspects of Public UCB Banking

As some authors have highlighted (Brown and Williams 2015), the exchange of unites inside the international circuit of public banks is not free and the export price may be higher than the cost of storage. The biological material circulates within a complex network where different actors are involved and where ethical, legal, technical, and economic considerations are necessarily intersected. Even if a system is based on unpaid and voluntary donations, the transfer of biological material among different countries implies a flow of money and therefore a bioeconomy issue arises (Petrini 2014). In the public banking system UCB has a monetary value but this does not automatically imply that it is a source of "profit" if the financial gain is spent on covering the operating cost. The particular kind of *biovalue* that UCB as a biological product assumes in different models of storage and distribution is a current issue.

The Informed Consent

Umbilical cord blood collection and storage raise several ethical and legal problems, some broadly debated in literature, as informed consent, ownership and patentability of UCB-derived products. The issue of cord blood ownership falls within the general discussion of the body ownership, widely debated from an ethical and legal point of view. Some consider the cord blood sample as a child property, since it belongs biologically, genetically, and developmentally to the child while others suggest that cord blood, once the cord is cut, is the mother's property. All national laws recognize as a legal person a child who is fully outside of the mother's body even though such an individual is not able to understand and provide consent. As well stressed by contemporary bioethics, informed consent is close to the principle of autonomy that is defined as "the quality or state of self-governing" (Petrini 2010). Therefore, the principle of autonomy could not be applied in case of cord blood storage since a human body part is used without the individual knowledge. From a practical point of view the main issues relative to the informed consent for cord blood storage can be summarized as follows: (1) who has to give the consent; (2) what content should be included; (3) who is going to accept it; (4) when consent should be given. The international community agrees that informed consent to donate umbilical cord blood for clinical or research use must be given by the newborn's mother. On the other hand, it is not so obvious to define what to do when the baby donor reaches the legal age of maturity. Legally, indeed, the right to dispose what it could be done with his/her body and its parts could be recognized to him or her.

In general the involvement of the father, besides that of the mother, would be recommendable; the content of the informed consent should be clear and exhaustive, including information about the potential uses of discarded units (i.e., the production of compounds or drugs for therapeutic or/and nontherapeutic purposes). A controversial situation arises when products developed from human biological material, donated for altruistic purposes, have the potentiality to be exploited commercially. As discussed by Petrini, the above mentioned possibility should be clearly and exhaustively disclosed during the informed-consent process and the donor should have the option to refuse consent (Petrini 2012).

Informed consent should always be obtained in a proper way and, since it needs time for reflection, should not be given just before childbirth that is without the possibility of adequate reflection on the part of the mother. Furthermore, it is important to communicate all medical circumstances of the mother or neonate that may prevent umbilical cord blood collection (American College of Obstetricians and Gynecologists 2015).

The National Academy of Medicine (US) Committee on establishing a National Cord Blood Stem Cell Bank Program, in the document "Cord Blood: Establishing a National Hematopoietic Stem Cell Bank Program," addresses issues and gives recommendations about the management of the informed consent (Meyer et al. 2005). The informed consent raises also the issue of data protection as underlined by the Belgian Bioethics Committee (2007).

Clamping and Cutting Timing of the Umbilical Cord at Birth

An issue related to the collection of cord blood cells is the clamping and cutting timing. At birth, if cord blood is not clamped, blood flow between baby and placenta may continue for several minutes. It provides all the necessary nutrient and blood to the unborn and that is why it is important for parents be certain that, in the donation procedure, clamping time will not harm the baby. The optimal timing of umbilical cord clamping is still being discussed today and the timing may vary worldwide according to clinical policy and practice. The WHO Guideline "Delayed umbilical cord clamping for improved maternal and infant health and nutrition outcomes" states "Delayed umbilical cord clamping (not earlier than 1 min after birth) is recommended for improved maternal and infant health and nutrition outcomes" and "Early umbilical cord clamping (less than 1 min after birth) is not recommended unless the neonate is asphyxiated and needs to be moved immediately for resuscitation." Immediate and long-term benefits of delayed umbilical cord clamping based on the results of randomized controlled trials and other type of studies have been reported for term, preterm/low birth weight infants and mothers, as summarized by the WHO Guideline (World Health Organization 2014). It is relevant that parents of the newborn donor receive complete information concerning the timing/procedure of cord blood clamping and, as suggested by the Study Group on Banks of Umbilical Cord Blood (Pontifical Academy for Life Banks of Umbilical Cord Blood 2013), the timing in seconds of the clamping should be registered on the clinical chart.

Prospective

Research, analyses and studies on the properties of cord blood hematopoietic progenitors and their clinical applications are attracting an increasing amount of attention, with the potential for even better results and new several indications for cord blood use. Methods are currently being investigated to improve the speed of engraftment and reduce transplant related mortality (e.g., the use of double cord blood transplants). Following the advancement of knowledge and the growing body of evidence that has emerged from research, UCB is being used as well for the treatment of nonhematopoietic disorders and immune modulation (Ballen et al. 2013). Furthermore these achievements have generated high expectations in the general population, with a growing number of clinical trials being designed to investigate UCB in different and various conditions, including Alzheimer's disease, autism, and diabetes mellitus type 1 and 2 (Mahla 2016; Roura et al. 2015). The popularity of cord blood has grown together with the specific market and the considerable appeal of the promises made in relation to cell therapy. Recently the possibility to use cord blood platelet and cord blood plasma for the preparation of Cord Blood Platelet Concentrate (CBPG) or Cord Blood Eye Drops (CBED) for clinical use has been also explored. Moreover studies have begun to investigate new reagents from cord blood units not suitable for transplant. All these new applications, in particular the development of non-therapeutic products with a commercial value raises various ethical and legal concerns. The most promising and innovative approach seems to be the one that uses nonhematopoietic stem cells from cord blood and placenta. Particularly, mesenchymal stem cells (MSCs) have been isolated from cord blood and placenta and are considered promising for therapeutic purposes (Roura et al. 2015). These findings have shown that UCB is a powerful biological resource and an important area in human regenerative medicine, which does not raise the ethical problems associated with procedures involving embryonic stem cells. The potential of UCB for regenerative medicine should be considered to be speculative until further evidence of possible benefits emerges and this aspect is also related to the sustainability of private banking: the best available research evidence will be indispensable to formulate future recommendations regarding autologous CB banking. As stated also by the American Academy of Pediatrics, clinical research advancements in the context of regenerative medicine might have an impact on cord blood banking in the future (Shearer et al. 2017). Equally, advertising potential future uses that are not supported by clinical evidence, for example a cure for heart disease or autism, should be discouraged or banned. As argued by many, it will undoubtedly be necessary to evaluate the sustainability of cord banking in the future according to any new evidence that emerges, including the actual clinical use. The annual activity survey of the European Society of Blood and Marrow Transplantation (EBMT) reported for 2015 a decreasing use of unrelated cord blood as a donor source for total HSCT, in contrast to a rise seen in haploidentical donor HSCT. The survey states that the decrease was observed for myeloid and lymphoid malignancies but not for nonmalignant disorders, where the use of unrelated cord blood is stable over time (Passweg et al. 2017). Currently a large clinical trial is ongoing with the aim to compare UCBT with haploidentical HCT (Ballen 2017).

Conclusions

Cord blood transplantation has been shown to be effective to treat patients with different and serious disorders and its use should be supported. As recommended by the scientific community, parents of the donor must receive adequate information about cord blood banking. In particular: (1) personal or familiar storage of cord blood is not a standard of care mainly because the stored cord blood may contain the same malignant cells that caused the disease; (2) direct cord blood banking should be encouraged only when it is known that in the family there are clinical conditions that could benefit from CB transplantation; (3) CB storage for personal use, as "biological insurance" for the newborn/family, should be discouraged since there are not scientific evidences to support autologous cord blood banking. Institutions and organizations, on the other hand, should (1) inform in a clear, correct, and not misleading way and communicate in a way that protects the public by "therapeutic illusions"; (2) should promote CB voluntary donation supporting the ethical value of a donation that is anonymous, voluntary, and nonprofit; (3) adequate policies should be developed for result disclosure, data protection, research strategies, and informed consent.

Acknowledgements *Competing interests*: the authors declare that they have no competing interests.

References

American College of Obstetricians and Gynecologists (2015). Umbilical cord blood banking. Committee opinion no. 648. Obstet Gynecol, 126(6), e127–e129. https://doi.org/10.1097/AOG.0000000000001212

Armitage S (2016) Cord blood banking standards: autologous versus altruistic. Front Med 2:94. https://doi.org/10.3389/fmed.2015.00094

Ballen K (2017) Update on umbilical cord blood transplantation. F1000 Res 6:1556. https://doi.org/10.12688/f1000research.11952.1

Ballen KK, Gluckman E, Broxmeyer HE (2013) Umbilical cord blood transplantation: the first 25 years and beyond. Blood 122(4):491–498. https://doi.org/10.1182/blood-2013-02-453175

Ballen KK, Verter F, Kurtzberg J (2015) Umbilical cord blood donation: public or private? Bone Marrow Transplant 50:1271–1278. https://doi.org/10.1038/bmt.2015.124

Belgian Advisory Committee on Bioethics (2007) Opinion no 42 of 16 April 2007 on umbilical cord blood banks. Retrieved from https://www.health.belgium.be/en/opinion-no-42-umbilical-cord-blood-banks. Accessed 28 August 2018

Brown N, Williams R (2015) Cord blood banking—bio-objects on the borderlands between community and immunity. Life Sci Soc Policy 11:11. https://doi.org/10.1186/s40504-015-0029-8

Gluckman E, Broxmeyer HA, Auerbach AD et al (1989) Hematopoietic reconstitution in a patient with Fanconi's anemia by means of umbilical-cord blood from an HLA-identical sibling. N Engl J Med 321(17):1174–1178. https://doi.org/10.1056/NEJM198910263211707

Kurtzberg J (2017) A history of cord blood banking and transplantation. Stem Cells Transl Med 6(5):1309–1311. https://doi.org/10.1002/sctm.17-0075

Mahla RS (2016) Stem cells applications in regenerative medicine and disease therapeutics. Int J Cell Biol 2016:6940283. https://doi.org/10.1155/2016/6940283
Meyer EA, Hanna K, Gebbie C (2005) Cord blood: establishing a National Hematopoietic Stem Cell Bank Program. The National Academies Press, Washington, DC. https://doi.org/10.17226/11269
NetCord-FACT (2015) NetCord-FACT international standards for cord blood collection, banking and release for administration accreditation manual, 6 edn. http://www.factwebsite.org/uploadedFiles/Standards/NetCord%20FACT%206th%20Ed%20Manual%20Draft.09.01.15.pdf
Passweg JR, Baldomero H, Bader P et al (2017) Use of haploidentical stem cell transplantation continues to increase: the 2015 European Society for Blood and Marrow Transplant activity survey report. Bone Marrow Transplant 52(6):811–817. https://doi.org/10.1038/bmt.2017.34
Petrini C (2010) Umbilical cord blood collection, storage and use: ethical issues. Blood Transfus 8(3):139–148. https://doi.org/10.2450/2010.0152-09
Petrini C (2012) Ethical and legal considerations regarding the ownership and commercial use of human biological materials and their derivatives. J Blood Med 3:87–96. https://doi.org/10.2147/JBM.S36134
Petrini C (2013) Ethical issues in umbilical cord blood banking: a comparative analysis of documents from national and international institutions. Transfusion 53(4):902–910. https://doi.org/10.1111/j.1537-2995.2012.03824.x
Petrini C (2014) Umbilical cord blood banking: from personal donation to international public registries to global bioeconomy. J Blood Med 5:87–97. https://doi.org/10.2147/JBM.S64090
Petrini C (2015) Ethical issues related to the collection, storage, and use of umbilical cord blood cells. In: Domen RE (ed) Ethical issues in transfusion medicine and cellular therapies. AABB Press, Bethesda, MA, pp 57–72
Pontifical Academy for Life Banks of Umbilical Cord Blood (2013) Pontifical Academy for Life, Rome, p.55
Roura S, Pujal JM, Gálvez-Montón C, Bayes-Genis A (2015) The role and potential of umbilical cord blood in an era of new therapies: a review. Stem Cell Res Ther 6:123. https://doi.org/10.1186/s13287-015-0113-2
Shearer WT, Lubin BH, Cairo MS et al (2017) Cord blood banking for potential future transplantation. Pediatrics 140(5):e20172695. https://doi.org/10.1542/peds.2017-2695
World Health Organization (2014) Delayed umbilical cord clamping for improved maternal and infant health and nutrition outcomes, guideline. http://www.who.int/nutrition/publications/guidelines/cord_clamping/en/. Accessed 28 August 2018

Index

A
Achilles tendon injury, 62
Acute kidney injury (AKI), 77
Acute liver failure, 23–24
Acute myocardial infarction (AMI), 23, 78
Adipose-derived mesenchymal stem cells (ADSCs)
  culture and expansion, 79
  isolation and characterization, 81–82
Adipose tissue, 79, 128
Adult stem cells, 5
Advanced DMEM medium, 38
Alzheimer's disease, 19, 150
American Academy of Pediatrics, 150
American Association of Blood Banks (AABB), 146
Angiogenesis, 22
Antibodies, 43
Apoptotic PBMC (Apo-PBMC), 20, 21
Argosomes, 61
Arrhythmia, 6
AutoXpress® Platform, 127

B
BD Mosaic™, 126
Beckman Coulter, 42
Belgian Bioethics Committee, 149
Bioethics, 6, 148
Biological agents, 106
Biological pharmaceuticals, 53
Bone marrow stromal stem cells, 120
Bone morphogenetic protein 4 (BMP4), 24
Bottom-loading method, 43
Bovine serum albumin (BSA), 38
Brain derived neurotrophic factor (BDNF), 21
British Virgin Health Bank, 145
Broxmeyer, 143

C
Cardiosphere-derived cells (CDC), 59
Cartistem, 125
CD34+ cells, 58, 108
Cell free pharmaceuticals, 28
Cell suspension culture, 105
Cell therapy, 7
Central nervous system (CNS), 5
Centrifugation, 40
Chronic kidney disease, 78
Chronic obstructive pulmonary disease (COPD), 124
Clamping and cutting timing, 149
Clinical trial study, 110
COPD assessment test (CAT) score, 124
Cord blood banking, 144
Cord Blood Eye Drops (CBED), 150
Cord Blood Platelet Concentrate (CBPG), 150
Cosmeceuticals, 64
Cosmetic products, 64
Crohn's disease, 125
Cryopreservation, 106
Cushion-based isolation method, 44

D
Dedicated storage, 144
Density gradient method, 43, 44
Differential centrifugation, 40, 42
Differentiation, 105, 132

P. V. Pham (ed.), *Stem Cell Drugs - A New Generation of Biopharmaceuticals*, Stem Cells in Clinical Applications, https://doi.org/10.1007/978-3-319-99328-7

E

Embryonic stem cells (ESCs), 4, 12, 19, 94
Endosomal sorting complexes required for transport (ESCRT)-I, 57
Endotoxins, 133
Engineered tissues/organs, 101
Enzymatic method, 129
European Marrow Donor Information System, 145
European Society of Blood and Marrow Transplantation (EBMT), 150
Exosome, 7
  biology, 55–58
  cargo, 57
  CDC, 59
  cGMP standpoint, 66
  characterization, 58
  cosmeceuticals, 64
  disease phenotypes, 70
  ESCRT machinery, 57
  EV therapeutics, 69
  exosome-based therapies, 54
  and exosome mimetics, 69
  FDA, 69
  health and safety risk, 63
  hiPSC-MSC, 63
  HSP70 and HSP90, 57
  liposomes/nanoparticles, 62
  literature, 53
  MSC, 62
  MSC-derived micro-vesicles, 58
  nano-vesicles, 66
  nomenclature, 55
  protein, 57
  repair and regeneration, 54
  RNA, 57
  rodent models, 57
  scalable production, 65–67
  signaling pathways, 60
  skin, 63
  STAT3 expression and phosphorylation, 63
  Wnt4, 63
Exosome biology, 55–58
Exosome nomenclature, 55
Extra-cellular matrix (ECM) components, 60
Extracellular vesicles (EVs)
  acetate buffer neutralizes, 45
  ADSCs, 83
  brain injuries, 78
  CD34 + cells, 11
  cell-cell communication, 77
  centrifugation, 40
  characterization and quantification, 80–81
  collection and processing, 37–40
  comparisons, 46
  cytokine concentration, 85
  differential centrifugation, 40
  FBS, 38
  FSB, 38
  IFN-gamma *vs*. TNF-alpha, 85
  ISEV, 38
  kidney disease, 77
  markers, 82
  MSC stress protocols, 38
  MSC-EVs, 78
  MSCs, 37, 39
  MSCs-derived EVs, 7
  pellet formation, 41
  production, 86
  protocols, 42
  quantification, 80
  quantity and type, 38
  rotor manufacturer, 41
  rotors, 41
  secretory processes, 87
  sources, 37
  storage, 46–47
  UC-MSCs, 83
  viscosity values, 40

F

Fetal bovine serum (FBS), 38, 125
Fetal calf serum (FCS), 39
Ficoll gradient centrifugation, 102
Fixed-angle rotor, 41
Flow cytometry data, 56

G

Graft-versus-host disease (GVHD), 78
Granulocyte-derived EVs, 47

H

Hematopoietic progenitor stem cells (HPSCs), 11
Hematopoietic stem cells (HSCs), 94
Holoclar products, 97
Human embryonic stem cell, 11
Human umbilical cord blood, 94

I

IFN-gamma, 83
Immunoaffinity isolation, 42

Immunosuppression/immunoregulation, 24
In vitro differentiation, 79–80
Infected cells, 39
Informed consent, 148, 149
International Society for Extracellular Vesicles (ISEV), 55
Iscove's Modified Dulbecco's Medium (IMDM), 127

K
Keloids, 63
Kit-based precipitation, 45–46

L
Limbal stem cell manufacturing, 98
Liposomes, 64

M
Magnetic assisted cell sorting (MACS), 102
Magnetic field exposure, 39
Mammalian cells, 6
Mature stem cells, 6
Mesenchymal stem cells (MSCs), 19, 150
  adipose tissue, 128
  BM-MSC isolation, , 126, 125
  bone marrow-derived, 125
  characteristics, 120, 121
  commercialized products, 127
  definition, 121
  differentiation, 121
  EVs, 80
  GVHD treatment, 123
  HLA class I antibodies, 124
  human clinical trial, 124
  IMDM, 128
  immune cell proliferation, 122
  immune modulation capacity, 122
  in countries, 123
  in vitro expansion, 133
  in vivo experiments, 120
  isolation, 124
  markers, 80, 120
  mechanisms, 122
  microcarrier-based culture system, 131
  MNCs, 125
  modulatist, 124
  phenotype, 81, 132
  products, 123
  PRP, 129
  quality and safety, 132–133
  quality control, 132–133
  scale-up systems, 130–131
  SVFs, 129
  T flasks, 129
  UC tissues, 128
  UCB-MSCs, 127
  UC-MSCs, 83
  umbilical cord blood-derived, 125
Microcarrier-based culture system, 131
Microvesicles, 7–12
Modulatist, 125
Mononuclear cells (MNCs), 125
Multi-vesicular bodies (MVB), 57
Mycoplasma, 39

N
National Bioethics Committee, 146
Neurodegenerative disease, 20
Neurogenesis, 21–22
Neuronal precursor cells, 4
Neuroprotective cells (NPCs), 5
Neurotropic factors composition analysis, 21

O
OptiPrep datasheet, 43
OptiPrep gradient, 43
Organ regeneration technologies, 60
OX40lg-expressing ADSCs, 109

P
Paracrine factors, 19, 26, 28, 29
Parkinson's disease, 5, 19, 21
Phenotypic analysis, 81
Plasma membrane, 61
Plastic vessel surfaces, 121
PrepaCyte®-CB, 127
Private cord blood banking, 145
Prochymal, 98, 123
Progenitor cell proteins, 70
Proteins, 19
Public *vs.* private banks, 147
Pure stem cell products, 95

Q
Quantum cell expansion system, 129

R
Regenerative medicine, 94

Regenerative therapy, 19
Regulatory requirements, 68

S

Scale-up cell expansion, 134
Scanning electron microscopy, 39
Secretome
  AD-MSC, 24
  ADSC, 21
  angiogenesis, 22
  Apo-PBMC, 21, 23
  BMC and PBL, 23
  BM-MSC, 21
  cardiac regeneration
    and cardio-protection, 22–23
  HUCPVC-MSCs, 21
  in vitro and in vivo studies, 21
  neurodegenerative disease, 20
  production, 28
  proteomic analysis, 21
  regenerative application, 28
  sources, 20
  visceral endoderm, 24
  wound healing, 25
Self-renewal, 132
Senescence, 132
Serum deprivation, 38
Size exclusion chromatography, 45
Size exclusion filter method, 44
Skin, 63
Skin wound healing model, 59
Statistical analysis, 81
Stem cell biology, 58–62
Stem cell products
  adherent cell culture, 105
  ADSCs, 107
  advantages and disadvantages, 95
  blood and fat, 95
  1st generation, 95, 97, 107
  2nd generation, 95–98
  3rd generation, 98–99, 109
  4th generation, 99–100, 108
  5th generation, 100
  holoclar products, 97
  HSC-rich fractions, 95
  HSCs, 94, 95
  iPSCs, 99
  isolation & enrichment, 102
  MNC transplantation, 107
  modification and differentiation, 105
  MSCs, 95, 97
  proliferation and expansion, 104–105
  purity level and characteristics, 94
  solid tissues, 102
  storage/conservation, 106
  SVFs, 107
  T2DM, 108
  technologies, 102–104
  transgenic T cells, 99
  treatment, 107
Stem cell therapy, 94
Stem cells, 4, 19
  endosomal compartment, 7
  ESCs, 4
  EVS, 7
  exosomes, 7
  hidden cells, 5
  limitations, 6–7
  microvesicles, 8–11
  physiological/laboratory conditions, 4
  proliferation and differentiation, 6
  properties, 4
  therapeutic use, 5–6
  types, 4
Stirred-tank bioreactor, 131
STRIMVELIS, 99
Stro-1, 120
Stromal vascular fractions (SVFs), 79, 107, 129
Sucrose solution, 43
Systematic experimentation, 60

T

Tendon, 62
Therapeutic approach, 5
Time equivalences, 42
TNF-alpha, 83
Top-loading devices, 43
Total Exosome Isolation Reagent Kit, 80
Transforming growth factor (TGF)-β, 38
Transmission electron microscopy, 81
Tumor cell-derived EVs, 11
Tumorigenicity, 133
Type 2 diabetes mellitus (T2DM), 107

U

Umbilical cord blood (UBC), 126
  bank operations, 144, 146
  bio-economic aspects, 148
  biological insurance, 151
  CB stem cells, 146
  cells biology, 144
  clinical applications, 146
  donation, 144

Europe, 145
history, 143–144
HSCT, 145
private cord blood banking, 145, 147
prospective, 150
regenerative medicine, 150
stem cells, 144
WHO guideline, 149
Umbilical cord blood collection and storage, 148
Umbilical cord blood transplant (UCBT), 144
Umbilical cord stem cells, 4
Umbilical cord-derived mesenchymal stem cells (UC-MSCs)
ADSCs, 78, 80, 82, 83
culture and expansion, 79
IFN-gamma, 83, 84
isolation and characterization, 82
TNF-alpha, 82
Urinary exosomes, 70
Urine-derived stem cell, 5

V
Viscosity values, 40
Vitamin K1, 64

W
Wharton's jelly, 5
World Marrow Donor Association (WMDA), 146
Wound healing models, 25–26, 63

X
Xeno-free media approach, 125